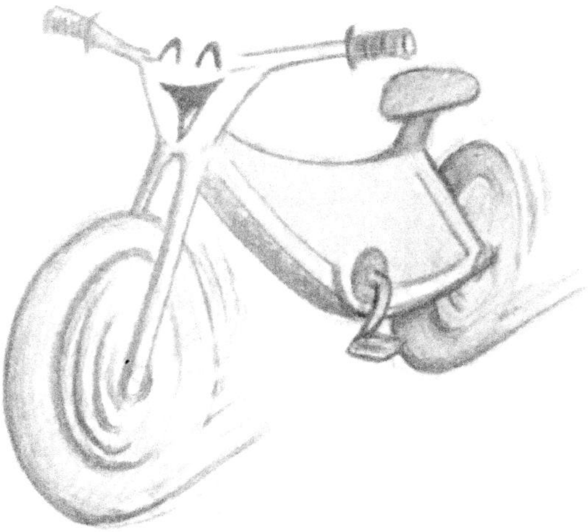

The Ultimate Do It Yourself Ebike Guide

By Micah Toll

Disclaimer: ebikes are fun. Actually, ebikes are awesome. But just like any vehicle traveling at high speeds, ebikes can be dangerous too. It is important that before attempting to build or ride any ebike, you first develop a working knowledge of the ebike and its components, as well as the safe operating practices and local rules of the road. Always wear a helmet. Always use lights when riding at night. Always assume other drivers are out to get you. Love your ebike but also treat it with the caution and respect that any vehicle capable of driving you off a cliff or into a wall duly deserves. Stay safe and have fun!

More legal mumbo-jumbo: I am in no way liable for any damage, death, injury to self or property that occurs as a result of this guide. You are entirely responsible for your own safety while building and operating any electric vehicle. This guide is for informational purposes only. If at any time you feel you are incapable of completing a build yourself, please seek professional assistance and do not attempt anything outside of your skill set.

Copyright © 2013-2017 by Micah Toll
All rights reserved.
No part of this book may be reproduced in any form or by any electronic or mechanical means including information storage and retrieval systems, without permission in writing from the author. The only exception is by a reviewer, who may quote short excerpts in a review. Printing is allowed only by the original purchaser of this ebook and only for use by the original purchaser. Printed copies may not be distributed or shared.

ISBN-978-0-9899067-9-1

Table of Contents

Foreword *9*

Chapter 1: Getting Started *14*

Chapter 2: Planning your Ebike *18*
Speed ---20
Range ---22
Weight---23
Charging time --24
Bicycle size--26
Bicycle features ---27
Accessories---30
Space---31

Chapter 3: Choosing your Components *34*
Bicycles--34
Motor--39
Batteries --54
Controller --68
Throttle ---73
Accessories---80
Endless Sphere---99
Ebikes.ca Hubmotor Simulator -----------------------100
Chapter 3 resources --------------------------------------102

Chapter 4: Purchasing Components *104*
Assembling your own set of components ------------120
Chapter 4 resources --------------------------------------130

Chapter 5: Installing the Kit *132*
Installing the motor--------------------------------------134
Installing the battery -----------------------------------145
Mounting the controller --------------------------------149
Installing the throttle -----------------------------------151

Other accessories -- 153

Chapter 6: Making the Connections 156
Soldering -- 159
Using connectors -- 164
Connecting everything to the controller ------------- 165
Battery connection --- 166
Throttle connection -- 167
Motor connection --- 169
Charger connector -- 175
Chapter 6 resources -- 177

Chapter 7: Finishing Touches and
Maintenance 179
Safety check -- 179
Test Ride --- 182
Maintenance -- 183

Chapter 8: Sample Ebike Builds 186
Charlie's commuter -- 187
Sally's slowpoke --- 190
Ron's racer -- 193

Acknowledgements 199

Foreword

Hello, Turbo Bob here. Through my bike blog, Facebook and YouTube pages, E-bike seminars, newspaper and radio interviews and magazine articles I am constantly getting out the word on all the great reasons to saddle-up on an E-bike. I have been riding, repairing and modifying bicycles since my pre-teenaged years. During those times I dreamed of motorized excitement and that led me to earn my nickname with countless turbocharged 4-wheel speed machines. Yet some 20 years back it came full-circle as my love of the bicycle was rekindled.

Now all that gas-guzzling is in my past, bikes have taken over my world and I couldn't be happier. My wife and I have been enjoying the fun, convenience and green savings (environmental and cash) of our E-bikes for over five years. Even with all the electric bikes I've ridden and reported on (almost 30 different company's E-bikes), I've seen that a well done conversion can easily put them to shame.

As you dive into this book, Micah Toll will unravel the mysteries of converting your own bike to the magic of electric-assist riding. With years of experience and hundreds of bike conversions under his mechanical belt, his expertise will allow you to not only to save time and money on your new source of transportation and fun, but allow you to tailor it exactly to your needs and desires. Sure, you can get a ready-made electric bike from one of many makers in the market place, yet the thrill of doing it yourself and having that one-of-a-kind machine that is "all you" trumps it hands down.

Micah's insight, informative writing style and learned knowledge will take you from the desire and pre-planning stages, through each step of organizing and obtaining the needed items, and get them installed and tuned for a bike that will change your life in the best of ways. The impressive list of suppliers and easy-to-understand technical information are just two of the in-depth ways this DIY book will take you from concept to completion. With your first ride on your newly converted electric-assist bicycle you will be singing tunes of praise for electric bikes and Micah's helpful guidance.

These pages go way beyond slapping a motor and battery on your bike for a way to flatten the hills and bring a smile to your face. Every aspect of the component choice and correct matching of them is covered. Plus sections on bike and accessory selection will give you the best chance possible for a successful and satisfying E-bike for years of riding. Micah has a way of making the semi-complicated procedures simple and even the needed math included is a breeze.

I've read other books on this subject and found this one to be a breath of fresh air when it comes to cutting through the long-winded, unneeded information you don't really want to decipher. Micah lays it out in everyday language and re-reading each paragraph dozens of times to find their true meaning is unnecessary. Of course like any book of this type you do want to go over it several times, but each round will bring enlightenment, not confusion.

One thing that is not overly stated in the text are all the reasons you might want and need an E-bike. The list is mind-boggling and each person will have their own major things about them they love. Micah and I share the love of electric bikes, as do so many others. Even though this concept was envisioned in the 1800's, it is truly a wave of the future that we can all ride together. With the influx of better motors, batteries and controllers, E-bikes are getting lighter and more efficient all the time. That is one more thing I like about this guide, it covers the newest and best the industry has to offer.

So join me, my wife, Micah and scores of happy E-bike riders. The time is now to use all an E-bike has to offer. E-bikes have added to our fleet of regular bikes in a way that makes long, tough rides fun and easy. I think you can follow my lead and make a bicycle your main form of transportation. Convert your bike and let's make our world a better, healthier and happier place to live. And if you are ever in San Diego, you are welcome to join our local twice-monthly E-bike club group rides. Hope to see you soon.

"Turbo" Bob Bandhauer

Creator and author of Turbo Bob's Bicycle Blog

Independent E-bike consultant and author

Life-long mechanic and technician

San Diego, California, USA

Chapter 1: Getting Started

By the act of beginning to read this book, you've already placed yourself on the right track to building your own electric bicycle. Congratulations, you're off to a great start! You're ready to enter the fun and fascinating world of building custom ebikes. Whether you've decided to build your own ebike to save money on gas, protect the environment, or just find a more convenient and fun way to get around town, by reading this guide you are taking the important first steps to acquire the skills and information that you'll need for a successful ebike build.

Before we jump ahead to all of the technical stuff, let's put things in perspective with a little background on motorized bicycles. These contraptions are not new. The first motorized bicycles were built almost 150 years ago using small gas engines adapted to regular, unmodified bicycles and actually predated motorcycles.

While primitive electric bicycles followed the first gas powered bicycles a few decades later, the first modern ebikes didn't start appearing until the early 1990's. These new ebikes debuted in Asia and became very popular in densely populated urban areas in China and India. However, these ebikes were primitive compared to the technologies available today. Most of these early ebikes employed lead acid or nickel cadmium batteries that were heavy and resulted in bikes with limited range and lengthy charging cycles.

Ebike technology improved slowly throughout the 1990's until the early 2000's with advances in nickel metal hydride and lithium ion rechargeable batteries. The new battery technologies reduced prices and facilitated the adoption of these higher quality batteries into electric bicycle design. As these more advanced batteries became more widely available, ebike adoption began expanding substantially. This expansion crept into Europe in the mid 2000's and more recently ebike use in North America has begun to accelerate considerably, especially on both coasts.

While commercially available ebikes started becoming more popular with the expansion of retail sale outlets, ebike enthusiasts all over the world quickly seized upon the opportunity to improve upon the commercial designs. While ebikes of various designs, sizes, and capabilities can be purchased around the world, most of the fastest, strongest and highest quality ebikes are custom made by enthusiasts in their basements and garages.

That's enough of a history lesson though. Let's move on to the present day. Because ebike laws vary from country to country, most ebike manufacturers have adapted their designs to conform with the fairly strict European Union (EU) statutes that limit ebike power to 250 watts and speed to 25 km/hr (15 mph). For most people, these limitations to speed and power are restrictive and don't provide for much more 'oomph' than a regular, non-electric bike. What fun is that??

Most standard bikes can easily be pedaled to reach speeds of over 25 km/hr (15 mph) and an average human can produce nearly 250 watts of pedal power. In fact, many professional cycling athletes can generate more than 400 watts of pedal power over a modest period of time. This simple comparison demonstrates just how low the 250 watt EU legal limit really is.

As if the power and speed limitations on most commercial ebikes weren't restrictive enough, the high cost of these ebikes is often the nail in the coffin. The least expensive of the commercially available ebikes sell for around $500, but these cheap ebikes usually use heavy, outdated batteries and weak, inefficient brushed motors. A few ebikes in this extreme-bottom bracket price range have begun using lithium batteries, but they also choose the cheapest kinds in order to keep cost down. This translates into a disappointing riding experience and often the unfortunate result of needing to pedal a heavy ebike home when your batteries have prematurely run out of juice. Also, these weak batteries need to replaced often, sometimes even once a year in the case of lead acid batteries, if you plan to get any serious use out of the ebike.

Ebikes using reasonably acceptable motors and battery technology start at around $1,200 retail and can escalate in price to well over $2,000 by the time you start getting into the better and more desirable component technology. Higher-end ebikes can go for over $5,000 and I won't even get started with what the "luxury" ebikes can cost. For these prices, you could build yourself an ebike with twice the range, speed and power, leaving enough cash remaining to add on some very cool accessories too!

With this simple review of cost versus technology and satisfaction, it quickly becomes obvious why building your own ebike is the only way to go if you want to get great performance at a deep discount. In addition, the process of building your own ebike will be personally rewarding and a lot of fun - not to mention that your ebike will be the envy of all your friends!

Chapter 2: Planning your Ebike

Before you can build your dream ebike, you need to develop a plan. Typically, when I assist people in planning for their first ebike, they begin with "I want to spend $_____ to build my ebike." To me, that's like telling the waiter how much money you want to spend on dinner before you've even seen what's on the menu!

Of course, understanding your budget is important so that you don't get carried away, but a better method than starting to plan with a price in your head is to start with a list of specifications and capabilities that you desire for your finished ebike. Then, by carefully selecting the types of components you will ultimately use, you can bring the price into the range that fits your budget while meeting your performance needs.

For example, a typical plan for an ebike might begin something like this: "I want a bike that I can ride in a comfortable and laid-back position, that can reach speeds of 20 mph (32 km/h), can go at least 20 miles (32 km) on a charge, and can be recharged in 3 to 4 hours." With a fairly complete picture of how you'd like your ebike to operate and perform, you can now start planning for the components to build the ebike that meets your needs and budget.

To continue a little further with the desired specifications from the example above, the upright riding style means you are probably looking at a beach cruiser or comfort city bike, possibly with front suspension in the latter case. The speed and range of 20 mph and 20 miles on a charge, respectively, mean that you should choose a motor in the 400-750 watts range and a battery of mid-range capacity. The charge time of 3 to 4 hours translates to choosing a battery charger with a higher watt rating, which allows faster charging. If some of these terms don't make much sense to you yet, don't worry. We will go over all of the details of ebike components in the next chapter. Soon you'll have all the information you need to make the necessary choices regarding these and other components for your own ebike.

Another essential component of planning your ebike is the need to start with a firm understanding of how you want to use your ebike. Do you plan to commute to work daily on your ebike or use it more for Sunday afternoon cruising to the local ice cream parlor or along the river walk path? Do you want to be able to carry a child (or two) on the bike or are you planning on tearing up the dirt on an off-road course? Do you intend to use your ebike as your main vehicle for the short term as stop-gap measure until you get a car, graduate from high school or college, or perhaps upgrade to a bigger and better ebike? Or do you need this ebike to last you for many years of faithful and dependable service? These are all important considerations and you'll see in Chapter 3 how the answers to these questions will affect your choice of specific components.

When planning your ebike, you'll want to list your specifications for the many important factors including speed, range, weight, charging time, size, bicycle features and accessories. Let's start with a fun one - speed!

SPEED

This is often the first consideration for most people as they begin to plan for their new ebike, and is obviously a key consideration. The speed of most do-it-yourself ebikes ranges from 15 to 25 mph (25-40 km/hr), although speeds in excess of 30 mph (50 km/hr) are certainly obtainable. Advanced ebike builders have even achieved record-setting speeds of over 70 mph (115 km/hr) through the use of heavily modified components, though this is difficult, expensive and extremely dangerous. However, it does serve to show what ebikes are capable of doing at the high end. Want to learn more about extreme ebike builders? Google "LiveForPhysics" and prepare to be amazed.

A good standard for first time ebike builders is to aim for a speed of 25 mph (40 km/hr) or less. Ultimately, the choice is up to you and should be defined by how you plan on using the ebike but this is a good starting place. For people who live in urban areas where they intend to spend most of their time riding on the sidewalks or in specially marked bike lanes to avoid traffic, 15 mph (25 km/h) is generally plenty. Going much faster than 15 mph on a sidewalk or bike lane can be dangerous, especially when pedestrians or other potential hazards pop up unexpectedly. For those who live in more rural or suburban areas, and/or plan on

> Quick Tip: Planning for a slightly slower speed on your first ebike doesn't only make it safer, it can also help keep the cost down too!

spending more time riding on the actual road among cars, faster speeds of 20-30 mph (32-48 kph) are usually preferred and required. This makes it easier to keep up with the pace of the traffic, plus it gets away from the dependency of hanging to the side of the road and puts you into an actual designated lane with the cars. In urban settings, it is often safer to drive in the lane with the other vehicles instead of on the side of the road. In this manner, you won't be getting passed as frequently by vehicles and you can reduce the chances of getting smacked by someone opening a car door on the side of the road.

One fact that you'll want to keep in mind is that the faster you go, the more battery power you'll use ,and as a direct result, the shorter your cruising range will be (or the bigger your battery will need to be in order to provide the same range). Ultimately, it's all about trade-offs, which is why having a good plan is essential. You will see this more as we continue along with our planning.

Range

This brings us to the next factor you'll want to think about when planning your ebike: range. While range is *by far* the single most critical limiting factor for the majority of electric vehicles, electric bicycles have the advantage of an onboard auxiliary power source: you! Now you may be thinking, "I'm building this ebike so that I don't have to pedal!", which is perfectly fine, but just remember that unlike an electric car, if you ever go too far without charging and run out of juice, you aren't going to be stuck calling a tow-truck; you can generally always manage to pedal to a nearby electrical outlet or charging station, if not all the way to work or to home.

To avoid that situation altogether, you'll need to plan for your ebike's battery to have sufficient capacity for your intended usage. When choosing a goal for range, I generally tell people to think about what their longest commute might be, and then add another 25% to that distance to allow for some safety cushion as well as the likelihood of some minor detours along the way. For example, if you know you want to be able to commute 8 miles to work from your home, then your round trip commute is 16 miles, plus another 25% (4 miles) equaling a desired range of 20 miles. Of course if you can recharge at work you are even better off!

There are two important considerations to remember when it comes to planning your desired range. First of all, it is important to understand that an ebike's range is greatly affected by outside factors including rider weight, terrain, wind conditions and other smaller yet still important factors. Climbing even moderate hills for sustained periods of time can really suck the juice out of your battery. If you live in a hilly area, this is an important consideration that you'll want to consider. Second, batteries will slowly lose their charging capacity over their useful lifetime. If your ebike had a range of 20 miles when you first built it, after a year its range will be something less than 20 miles. Exactly how much less depends on the type of battery you choose and how often you use it. Both of these issues will be covered in more depth in the battery section found in Chapter 3.

Weight

Next, you need to seriously consider how much the weight of your ebike matters to you. Ebikes are of course heavier than regular bikes, sometimes much heavier, depending on components. Motors and batteries are heavy components and most ebikes weigh anywhere from 40-75 lbs (18-35 kg). However, there are strategies that can be employed as part of the design process to minimize the weight of the bike, though this usually comes at the cost of some decrease in performance, increase in price, or both.

A heavy ebike is not necessarily a bad thing, especially if you are careful to spread the weight out along the bike and keep the weight as low to the ground as possible to lower the center of gravity. Again, think about how you'll be using your ebike. Do you need to carry it up a flight of stairs to bring it into your home or place of work, or do you plan on keeping it in a garage or elsewhere on ground level? Obviously, a plan for an ebike that needs to be lifted and carried should consider weight minimization. But if the ebike will always remain on ground level, or you can take it directly inside of an elevator, weight becomes much less of an issue. The more important consideration in planning for an ebike that is not weight limited is to spread the weight out evenly along the length of the ebike so it isn't loaded mainly at one end of the bike. You should also try to keep the weight down towards the ground as much as possible to maintain a low center of gravity.

One last thing on planning for the weight of your ebike: while it is true that reducing weight in electric vehicles helps increase their effective range, this has less of an effect on electric bicycles than on electric cars or motorcycles. Consider it. With ebikes, the rider greatly outweighs the ebike itself. Therefore, while reducing a total of 20 pounds from a 60 pound ebike would be a huge weight savings of 33%, if you factor in a 170 pound rider, suddenly the total weight savings of the vehicle plus the rider drops from 33% to less than 9%.

Charging time

The charging time of your bike is a very simple element to control, but make sure to plan for what you need. While it would be nice to be able to charge a dead battery in just an hour or two, it is both cheaper and much better for the longevity of the battery to charge it over a longer time period, such as 6-8 hours or even overnight. If you plan on using your ebike during the day for traveling within its maximum range on a single charge, then charging slowly at night is the best option. If you plan to be driving your bike all day, such as from client to client or using it for deliveries, then a fast charger that can help replenish the battery whenever you have a few minutes to spare becomes the better option.

> Quick Tip: When driving in a straight line at constant speed, the weight of your bike won't make much difference, but for people who will spend a lot of time accelerating and maneuvering, like off-road riders, the total weight and distribution is more important.

Because the rate of charging is almost entirely controlled by the charger and is independent of the actual ebike, it is possible to maintain two chargers, one for highly beneficial slow charging at night and the other for fast charging during the day. We'll cover chargers in greater detail in Chapter 3.

Bicycle Size

The size of your ebike is your next consideration. The two main issues related to the bicycle's size are the wheel size (diameter) and the overall frame size. Smaller wheels usually mean a smaller bike which can help with carrying your ebike or fitting your ebike in a car if necessary, but smaller wheels have downsides as well. Smaller wheels mean that road surface irregularities such as pot-holes, bumps and street curbs as well as the occasional piece of street debris feel larger and more uncomfortable. Big wheels roll over obstructions more easily. Also, the same motor powering a smaller wheel will result in a slower top speed. Most motors have a fixed maximum rotational speed, so a smaller wheel will travel less distance in the same number of rotations than a larger wheel, even when they both spin at the same rotational speed. If it sounds confusing, think about rolling a basketball and a baseball one revolution each - which goes further?

Choosing a frame size is important because your personal comfort on the bike is essential. You'll want to try out a few bikes to make sure that your arms feel comfortable with the distance to the handlebars and that your feet easily reach the pedals. Most bikes are designed so that when they have been properly adjusted for you, your feet don't actually sit flat on the ground. This is done on purpose to make sure that there is an appropriate amount of bend to the knee during the pedaling stroke. Because most electric bikes get pedaled very little, you may be more comfortable lowering the seat a bit to the point that your feet sit flat on the floor, even though this isn't the 'standard practice' for good pedal fitment.

Additionally, some bikes are designed with what is called a "pedal forward" or "crank forward" design, where the pedals are actually mounted further to the front of the bike than is considered standard, allowing the rider to maintain a proper bend in the knee while still being able to place his or her feet on the ground. These bikes are usually of the cruiser or comfort bicycle style.

Planning the size and style of the bicycle that you intend to use is an important step to ensure that your ebike is as comfortable as possible. Don't rush this step; take the time you need to make sure your bicycle feels right. Hopefully the two of you will be spending a lot of pleasurable time together.

Bicycle features

Bicycle features go hand-in-hand with the frame size when it comes to matching a bicycle with your comfort requirements. There are quite literally thousands of bicycle designs to choose from and you'll want to make sure you plan out what features you are looking for to help narrow down the array of options. For example, if portability is an important requirement for you, a folding bicycle or a bicycle with a quick release wheel will make it easier to put the bike into a car.

Many people desire suspension in a bicycle. This means that there is some type of spring or shock absorber built in to the frame to cushion the small irregularities of the road surface. Front suspension is achieved through a suspension fork, while rear suspension usually means the rear half of the frame is a separate piece, connected with at least one hinge and suspension unit.

Front suspension is relatively easy to find on bicycles and doesn't add a significant amount of cost or weight, though you should be wary of the least expensive of front suspension bicycles because the components are usually poor quality. A cheap suspension fork is actually a much worse deal in the long-run than a cheap non-suspension fork.

Also, you should know that a front suspension fork is not designed to handle the stresses of a hubmotor and should only be used with small motors in the range of 250 to 500 watts. A motor larger than 500 watts can occasionally break the fork by destroying the dropouts (the part of the fork that holds the motor's axle). This is a very bad thing! Unless you can find a replacement fork, which will be more or less difficult depending on the type of bike, this often means replacing the entire bike. This issue can mitigated though by using a torque arm, which we'll cover later on.

Rear suspension is less common and usually limited to better quality mountain bikes. This component, while nice, will add some additional weight to the bike as well as take up space within the frame, reducing some of the options for battery placement. Just like with front suspension bikes, a cheap rear suspension bike is usually worse than a cheap non-suspension bike. Again, it's all about tradeoffs.

You also need to consider other user-specific components, such as how many gears you may want on your ebike and if you actually think you'll be needing to shifting gears at all. Many electric bikes spend their entire lives in the highest gear to facilitate pedaling when the bike is at full speed. Ebike motors have plenty of torque so pedal assist at start off is typically not needed or desired. Also, ebikes accelerate fairly quickly meaning it is often easiest to just leave the bike in highest gear instead of shifting through the gears in rapid succession. The only time this isn't true is for the weakest of motors, usually in the 250 watt range, and for heavy riders that may bog down the motor a bit more. In these cases, it may be helpful to add a little pedal power in the beginning while the bike is in low gear.

Disc brakes are usually a good addition to electric bikes that travel faster than 20-25 mph (32-40 km/hr), but the traditional and highly functional rim brakes are still acceptable for most electric bikes. Just know that you'll need to change your brake pads more often than on a

> Quick Tip: If there is one place you don't want to scrimp, it's on your brakes. Spend a little extra money on good brakes and pads. Your safety is certainly worth it.

standard bicycle since you'll be going faster and using the brakes more as compared to pedal power cruising alone. Again, quality is important here. Good quality rim brakes will still be better than cheap disc brakes.

Accessories

With the basics already considered, you now need to think about any of the accessories that you might want to include on your ebike. This isn't as crucial as choosing the right bicycle or motor, but it will still be important to making sure you are as happy with your ebike as possible.

For anyone living in an urban area, you'll almost certainly want a good strong bell or horn to warn pedestrians and other cyclists as you approach from behind on sidewalks or shared walking/bike paths. Good horns make good neighbors! Well, maybe not, but they sure do make neighbors that stay alive!

A front and rear light is a must, though additional lights mounted on the frame or the wheel spokes are also good practice in order to better guarantee that other motorists see you on the road at night. If you plan on riding in poorly lit or off-road areas, you'll want more than a basic bicycle light. Options for stronger lights are provided in list form in Chapter 3.

You may want to consider package storage options for inclusion on your ebike if you plan to use it for commuting purposes. While much of what you may need on a daily basis can fit in a backpack, having a basket, rack or bicycle bag can be a nice addition for stashing extra items or storing a charger or extension cord onboard.

A sturdy kickstand is a must for just about any ebike, as you don't want it to fall over and potentially damage important and expensive components. A strong kickstand is even more important if you plan to have a large battery pack or have it mounted high up on a bicycle rack over the rear wheel. Consider a double kickstand, one with an extension on both sides of the frame, like the type used by scooters and mopeds. These are widely available in many bike shops or online and are a great, low-cost investment in the longevity of your ebike. Just make sure you get the right size to match your bike -- they are usually marked according to wheel size. A heavy ebike or one with batteries mounted high up can really benefit from a double kickstand.

Planning out your accessories ahead of time will ensure that you have available room on your ebike to install them all and don't wind up having to 'jury-rig' these devices after you've already completed your build.

Space

Finally, when planning your ebike build, make sure you plan to have the space and tools needed for the job. Everyone's work requirements are different; I've seen ebikes built in garages, backyards, kitchens and living rooms, so it really just depends on what space you have available (or as is sometimes the case for me, how long your wife lets you leave your bikes and components in that space).

Make sure that your work/assembly environment is clean and you have the space to lay out all of your parts to help make the build process as simple and straightforward as possible. In my experience, I've noticed that these projects tend to expand to fill the allotted time and space, so plan accordingly.

This would make a great work area...

...but this will work too, as long as you get permission from your significant other first!

Chapter 3: Choosing your Components

When it comes to building your ebike, there are really only five required components: the bicycle, motor, battery, controller and throttle. Without any one of those five parts, your ebike simply won't happen. After those five parts are chosen, everything else is just an accessory to customize your ebike for your specific needs and desires.

BICYCLES

As mentioned in the last chapter, choosing the right bicycle to start with is important for ensuring that your ebike is comfortable and fits you well. If you already have a bicycle that you enjoy and are ready to electrify it, then you're all set. If not, you'll need to begin the search for the right bike for you.

A great place to start is your local bike shop. Talk to them about what you are looking for in terms of size, shape and features. They can help you find a bike that fits well, but it will be up to you to make sure it is appropriate for an electric conversion (most bike shop employees are unfamiliar with electric bicycles and may even scoff at the idea of you taking a 'perfectly good' bicycle and making it electric - don't be deterred!).

A big bike can be fun, but be sure it's comfortable before you make the investment

There are certain items that you'll want to look for in a bike to make sure it is appropriate for electric conversion. First is a strong frame with strong dropouts, the place where the bike holds the wheels' axles. The best bikes for electric conversion are steel bikes with beefy steel dropouts. Aluminum bikes are acceptable, but will usually require an additional torque arm to strengthen the dropouts (we'll cover that in Chapter 5). Ironically, less expensive steel bikes often have stronger dropouts than some more expensive aluminum bikes, therefore this is one feature that you can't necessarily judge purely by the price.

You'll also want to pay special attention to the brakes. Brakes on an electric bike are much more important than a standard pedal bicycle because you'll be traveling at higher speeds most of the time. Think of it as the operational equivalent of riding a normal bike downhill all the time.

There are two main types of brakes, V-style rim brakes and disc brakes. Disc brakes generally require less frequent maintenance and provide stronger braking force, but come with their own downsides as well. They can be harder to change when the pads are worn down and can also complicate the installation of a hubmotor.

Some bicycles have disc brakes in the front and rim brakes in the rear. This is a very good compromise for electric bikes because it puts the stronger brakes up front where more braking force is needed, but leaves the rear wheel, which is more often converted to a hubmotor, unencumbered.

V-style rim brakes

Some bicycles have disc brakes in the front and rim brakes in the rear. This is a very good compromise for electric bikes because it puts the stronger brakes up front where more braking force is needed but leaves the rear wheel, which is more often converted to a hubmotor, unencumbered.

Disc brakes

If you are buying a new bike, your brakes should already be dialed in for maximum performance. If not, take it back to the store and ask them to adjust the brakes to your satisfaction. If you are using an older bike that you already own or are buying a used one, then you'll want to either tune them yourself or take them to a bike shop to have them checked out. Learning to do your own brake adjustments will be an important skill to have though once you start riding your ebike frequently. Check the resources section of this chapter for a few video links on adjusting your brakes.

If you choose a bicycle with suspension, make sure you understand how it works. As noted above, cheap suspension forks are often worse than a non-suspension fork, so keep that in mind. Higher quality and better designed suspension forks will have a knob on the top of the fork to adjust the damping and/or travel of the suspension. Rear suspensions will also be adjustable, though the mechanism will differ from bicycle to bicycle. You'll need to check the manual for your specific bicycle.

If this is your first ebike, you may be better off starting with a bicycle without suspension as it will simplify the build process and be less expensive.

The most important thing to remember is that whatever bike you choose, make sure it has a strong frame and beefy dropouts. A $150 steel frame department store bicycle can actually be more appropriate for electric conversion than a $1,500 lightweight aluminum or carbon fiber bike. If you can picture someone in spandex pedaling the bike in a race pack, then it's probably not the best bike to electrify.

Motor

Currently, there are three common types of motor drives used in electric bicycle conversions: hubmotors, mid-drive motors, and friction drive motors.

Hubmotors

At present, hubmotors are by far the most common motors used for electric bicycle conversions. Hubmotors are essentially a motor with its outer case built in the shape of a bicycle hub. A bicycle wheel is laced to the motor and the entire unit is swapped in place of the normal bicycle wheel.

Hubmotors are the simplest form of motor to use in an ebike conversion but are limited to a single gear ratio. This means that while they work well on flat ground, they can bog down on long or steep hills, especially when using a low powered hubmotor in the range of 250 to 350 watts.

Hubmotors	
Pros	**Cons**
Easy to install	Heavy
Less noise	Only one gear ratio
No maintenance	Expensive

Mid-drive motors

Mid-drive motors are usually mounted within the main frame triangle of the bicycle, or just below the pedal crank. They function by applying power to the standard bicycle chain to turn the rear wheel. The advantage of a mid-drive motor is that it allows the bicycle to use all of its existing gears along with the motor power, thereby letting you shift gears as in a manual transmission car. You can still use your gears with hubmotors, but they won't affect the motor. With a mid-drive, because the motor is tied into your chain drive, shifting gears also affects the motor's equivalent output. With this setup you can actively select the most appropriate gear for the immediate conditions, such as a lower gear for hills and a higher gear for flats.

Mid-drive systems aren't without their disadvantages though, which include increased maintenance of the chain/shifters and chance of chain breaking, increased dirt and grime build-up in the drivetrain, louder operation and somewhat more difficult installation. Also, they create a chance for a single point failure completely disabling the driving ability of the bike. Because the electric drive and pedal systems are no longer separate, a failure in the pedal drive system (such as a chain breaking) will render the electric drive useless, unlike hubmotors which can still drive even without a functional pedal drive system. While this case is rare, it can and does happen.

It used to be the case that there were only a few commercially available mid-drive motors units, and they were generally of poor quality and required changing many of the bicycle's main components. They also used old style brushed motors which are inefficient and limit your options for compatible components. The better mid-drive systems were therefore often one-off units that were custom made by enthusiasts using radio control (RC) aircraft motors and required significant design and machining experience in order to perform correctly.

However, there have been some big companies making promising advancements in the technology. The introduction of these newer, purpose-made ebike mid-drive kits has greatly increased the popularity of the mid-drive option. They are still greatly outnumbered by hubmotors, but mid-drive numbers are growing among a dedicated group of mid-drive enthusiasts.

The mid-drive kit I recommend most right now is the Bafang BBS01 (250-350 watts) or BBS02 (500-750 watts). This is a well designed kit that isn't terribly complicated to install, and actually looks great on many bicycles. While initial versions suffered from an underpowered controller and led to frequent overheating issues, the company has since solved this issue with larger controllers using better quality components.

Bafang has recently introduced another even more powerful mid-drive kit, the BBSHD, capable of over 1,000 watts of power. This is also a great mid-drive kit, though it is for riders who want higher power levels and it most suited for speeds of 25+ mph or off-road use. If you don't need to go much faster than 20 mph, stick with the BBS01 or BBS02.

There are mid-drive kits out there that are badged as E-rad kits. These are simply Bafang BBS01, BBS02 and BBSHD kits that have been rebadged to sell them under another name.

There are also mid-drive kits available from a company known as Cyclone. They specialize in higher power kits though that usually aren't appropriate for first time builders. If you want an insanely powerful, 3,000W mid-drive kit though, Cyclone might be a good option for you.

Mid-drive	
Pros	**Cons**
Multiple gear ratios	Complicated
Cheaper	Harder to install
Customizable	Loud

Friction drive motors

Friction drives work by mounting the motor directly to a shaft that presses against a bicycle's tire and drives the wheel using friction between the shaft and tire. Friction drive systems have the advantage of being simple and inexpensive, but most require frequent maintenance, including regular replacement of the bicycle tire being driven due to wear from the friction of the powered input shaft.

Friction Drive	
Pros	**Cons**
Cheapest	Wears out tires quickly
Relatively simple	Lower power/ inefficient

I recommend that any first time ebike builder start with a hubmotor because of the combination of simple installation, high reliability and nearly maintenance-free use. If you absolutely have your heart set on a mid-drive, check out the resources section at the end of this chapter for some helpful links with more information. There's a great installation video from California-ebike.com that I recommend checking out. Also, most of the controller, battery, throttle and other information in the rest of this guide will still apply to those drives.

The majority of hubmotors available today are known as "brushless DC hubmotors". Brushless means that instead of mechanical brushes dragging inside the motor to control timing and polarity, a separate unit known as an Electronic Speed Controller (often referred to as just "the controller") accomplishes the task more efficiently. DC stands for Direct Current and means that current is flowing in one direction around a circuit, as in the simple case of connecting a battery to a lightbulb like in a regular flashlight. Brushed hubmotors are getting harder to find because brushless motors have all but replaced them across the ebike industry. I recommend that you stay away from brushed hubmotors because they are an older technology that is quickly becoming a dinosaur in the ebike community.

There are two main types of hubmotors currently on the market, geared and direct drive motors. Direct drive hubmotors (sometimes called gearless) are the simpler of the two, so we'll start with those.

In a direct drive hubmotor, the axle that passes through the center of the motor is actually the axle of the motor itself, with the copper windings fixed to the axle. This whole axle assembly is called the "stator". The magnets are mounted in the outer shell of the hubmotor so that when electricity is applied to the stator, inducing a magnetic field and causing the magnets to move, the entire shell of the motor turns.

Gears from a Bafang BPM hubmotor

Geared hubmotors, on the other hand, have their cases connected to the stator through a gear reduction system so that for every rotation of the case, the motor inside actually turns many times faster. This allows the motor to work at a higher and more efficient speed, while still allowing the wheel to spin at a comparatively slower speed appropriate for road use.

Geared hubmotors are also smaller and lighter than direct drive motors, but are less powerful and can wear out more quickly. Most geared hubmotors are only rated for up to 350 watts of power, though the larger Bafang BPM geared hubmotor has a rating up to 500 watts and has been successfully used on ebikes at levels over 1,000 watts, though with a shorter life expectancy of the motor.

Choosing between a geared or direct drive hubmotor usually comes down to two considerations: your power and weight requirements. If you are designing your ebike with a lightweight setup in mind, you are pretty much limited to a geared hubmotor. If you are designing for a powerful ebike, especially one over 1,000 watts, your only hubmotor option is a direct drive hubmotor.

Choosing how much power you need in an electric bike can be a little confusing at first, especially because motor manufacturers use differing and often spurious methods for rating motor power. In the controller section we'll talk about calculating the exact power of your ebike. For now though, here are some rough guidelines to help you determine your power requirements. When traveling on flat ground, a 350 watt motor can easily achieve speeds of 20 mph (32 km/h) while the same motor traveling up a moderate 5% grade hill would get bogged down to around 8 mph (20 km/h). A 1,000 watt motor could achieve the same 20 mph (32 km/h) speed on flat ground but sustain approximately 15 mph (25 km/h) on a moderate 5% grade hill. The main advantage of a higher watt system is increased hill-climbing speed and power, as well as greater off the line (starting from rest) acceleration. Most people end up in the 500 watt neighborhood on their first ebike build, because this is a good middle ground that provides exhilarating yet controllable power without getting too expensive.

As you saw in the example above, a 350 watt motor and a 1,000 watt motor can both push your ebike to approximately 20 mph (32 km/h) on flat ground. This may seem counter-intuitive at first, but remember that the wattage is a rating of power, not speed. The power is what gets you going from a stop or propels you up a hill. To achieve more speed you have to either choose a motor with a higher RPM rating or increase the voltage of your system, or both.

Motors are often listed with a voltage rating as well as an RPM value. If a manufacturer's motor is listed as spinning 328 rpm at 48V, but you instead connect it to a 36V battery, you'll see a decrease in speed roughly equivalent to $36 \div 48 = 0.75$, meaning the new speed will be approximately 75% of the original speed. Alternatively, if the motor is listed as 36V and you run it at 48V, you'll see an increase in speed of $48 \div 36 = 1.33$ or a new speed equal to approximately 133% of the original speed rating of the motor.

If you are planning your ebike to attain a certain speed, it is important to make sure you know the RPM of the motor you intend to buy (or the rated speed) for a specific voltage level. If you accidentally buy a motor intended for 48V and run it at 24V, you'll be sorely disappointed to see it only drives half as fast as you had hoped.

Here's a tip that I use to determine how fast any motor will drive any bike. It all depends on the size of the wheel. There is a fairly long mathematical formula to calculate the final speed, but I've simplified it as much as possible by canceling out units and combining terms to result in what I call the "Micah Speed

Formula" (I've always wanted a formula named after me). It goes like this:

$$\text{eBike speed in mph} = \left(\text{Motor RPM}\right) \times \left(\text{Tire diameter in inches}\right) \times \left(0.003\right)$$

With this formula, all you need to do is multiply the motor speed in RPM (revolutions per minute) by the tire's diameter (in inches) and then multiply that number by 0.003. For example, most ebikes will use either a 26 inch or 20 inch tire. So if the motor states 328 RPM at a given voltage, to determine the speed on a 26 inch wheel, multiply 328 x 26 x 0.003 which gives a result of 25.5 mph (41 km/h). Not bad.

The above formula gives you a final speed in miles per hour. If you want the results in km/h instead of mph, you'll want to use this altered formula:

$$\text{eBike speed in km/hr} = \left(\text{Motor RPM}\right) \times \left(\text{Tire diameter in inches}\right) \times \left(0.0048\right)$$

You'll notice the tire diameter is still in inches (english units) while the speed is in km/hr (metric units). This isn't a mistake. Even in countries that use the metric system, bicycle tire diameters are still generally referred to in english units, such as a 26" tire.

Also, you can combine this formula with the one mentioned on the previous page for determining motor speed when changing voltages. For example, if we were using the 328 RPM motor identified above that we determined to give 25.5 mph (41 km/h) on a 26 inch wheel, but we wanted to change the voltage, we would need to recalculate. If the motor was listed as a "48V 328 RPM motor" and we wanted to use it on 36V instead, we would take the final speed we found and multiply it by 36/48 like this: 25.5 mph x 36 ÷ 48 = 19 mph (30 km/h).

Lastly, you should know that the faster you go, the greater the performance losses will you'll experience will be. This is mainly due to wind resistance, as well as a few other smaller factors and inefficiencies. For this reason, I usually subtract about 10% from the speed calculation derived from this formula to determine my "real world" speed after inefficiencies and losses due to the many environmental factors. If you're going really fast, like above 35 mph (56 km/h) then you might want to add a bit more than a 10% speed penalty.

Now, as if there weren't already enough options when it comes to hubmotors, you'll also have to choose if you want a front or rear hubmotor. Both have their (you guessed it) advantages and disadvantages.

A front hubmotor helps distribute the weight of the bike along its length which generally improves handling, especially if the battery is mounted on a rack above the rear wheel. It also makes fixing a flat tire easier since most flats occur in the rear tire where more weight increases the likelihood of road debris stirred up by the front tire actually puncturing the rear tire.

The disadvantage of front hubmotors is that the more powerful they are, the more likely you are to "burn rubber" when accelerating from rest, which is the situation where your tire loses traction and spins freely against the ground until traction is regained. Less weight on the front wheel means less traction (grip) from the tire on the road.

Rear hubmotors inherently have better traction due to more weight sitting over the rear of the bike. The better traction provided by rear hubmotors makes acceleration more comfortable and controllable.

Generally speaking, hubmotors of 500W or more are usually better in the rear because the higher power makes front drive ebikes unwieldy when accelerating from rest. Motors of less than 500W can be placed in the front or rear, and often the front is chosen for the weight distribution advantage.

Ultimately, the choice is a matter of comfort and is entirely up to you. When you're cruising in a straight line at top speed, which is where you'll likely spend 95% of your time on an ebike, the difference between a front versus a rear motor is essentially indistinguishable.

Front Vs. Rear Hubmotor	
Better weight distribution	Better traction
Fewer flat tires in motor wheel	Safely applies more power
Easier to install	Smaller motor can hide behind gears

There are many manufacturers of hubmotors, both geared and direct drive. Some common motors are listed below with their generally adopted naming schemes. Because most, if not all of these motors come from China, they sometimes pop up with different naming schemes for the same motors.

Nine Continent (sometimes abbreviated 9C) motors are powerful, rugged motors often used by ebike enthusiasts modding for extreme power and speeds. The motors are most commonly seen in a black case, though silver versions do exist. The motors weigh about 13 lbs (6 kg) and are available in several speed versions. The motors are commonly used with both 36V and 48V batteries and can handle up to approximately 2,000 watts before modifications are needed to assist with motor cooling issues.

Nine Continent

Crystalyte makes some of the strongest direct drive hubmotors available for bicycle conversions, but they are also some of the heaviest. Crystalyte motors are mostly suited for those trying to push the limits on how fast and powerful ebikes can be. A beginner can be very happy with a Crystalyte motor but should know ahead of time that this is a very heavy duty motor meant for use well over 1,000 watts. Similar motors include the QS motor and the Leaf Motor, both of which are direct drive hubmotors that can also handle power levels well over 1,000 watts.

The MagicPie by Golden Motor is a special type of hubmotor that incorporates a built in controller to make installation as simple as possible.

Magic Pie

The motor is quite strong but also one of the biggest and heaviest hubmotors available.

The MagicPie kits also come with everything you need to build your ebike, except for the battery, which means that installing it is very easy because all of the connectors are designed to mate with each other. Buying a MagicPie kit directly from the manufacturer means you can also pair it with an appropriate battery, but we'll talk about that more in the vendors section of Chapter 4.

Bafang BPM geared hubmotors are some of the strongest geared hubmotors available. They can't compete for power with direct drive hubmotors, but they will still blow away nearly any other geared hubmotor. They are usually marked as 500W but can safely handle up to 1,000 watts of power. Larger gears, magnets and windings make this a great motor for those looking to push their ebike to higher power levels but still want to use a lighter weight motor or retain some stealthy look to their ebike. MAC motors are in the same class as Bafang BPM motors - meaning

Bafang

they are beefier geared motors that are nearly as capable as the stronger class of gearless direct drive motors.

Bafang SWKX/SWKX5/SWXU/etc are smaller geared hubmotors made by the same company as the Bafang BPM. These motors are smaller than their big brother, but still quite capable. They are more appropriate for the 250W - 500W range.

Cute motors (that's the name, really) are very similar to the Bafang SWKX/SWKX5/SWXU/etc style motors in appearance and construction. Like the Bafangs, these motors are commonly used in 250W-500W applications. They are incredibly light for a hubmotor, weighing in at around 4 lbs (less than 2 kg). They are normally sold under model names such as Q85, Q100 and Q100H.

No name/eBay motors are available from many sources, obviously including eBay. These are hubmotors that are unmarked or unbranded but usually have characteristics similar to more well known hubmotors.

Sometimes these motors are available for significantly less than other well known hubmotors but this is strictly a 'buyer beware' situation. You may get a great motor that lasts years or you may get an inferior product that works for a month (or never works at all).

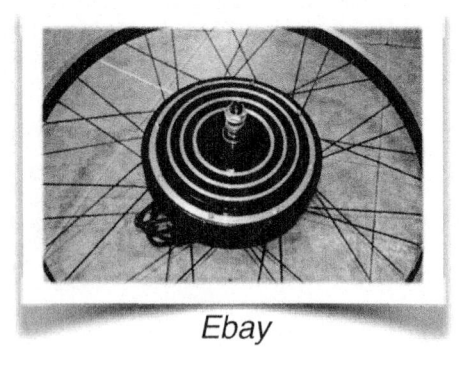

Ebay

Vendors will rarely publish specifications for motors like these so the actual final speed may be a mystery until you install them on your bike and discover it first hand.

There are truly great ebike kit deals on eBay, but you should always check seller feedback before jumping at what looks like a real bargain. Even so, I rarely recommend using a kit sourced from eBay unless you have previous knowledge, such as a recommendation from a friend who bought from the same vendor, to attest to the quality of the kit in question and the customer satisfaction policy of the vendor.

At the end of the day, there are likely many different motors that you could be quite happy with on your own ebike. The most important thing when choosing any motor is to plan for what speed you'd like to achieve. By choosing the right RPM and voltage combination for the motor, you can narrow in on your intended speed with many different styles of motors.

BATTERIES

Before we can even begin talking about different styles and chemistries of batteries, there are some basics that you'll need to understand. Some of this background information may already be familiar to you, which immediately gives you a leg up. For everyone else, pay special attention here.

Batteries are chemical energy storage devices that are used to supply DC electricity to a circuit. There are two main metrics used to define the specifications of a battery: voltage and capacity. Voltage is measured in volts (V) and capacity is measured in amp-hours (AH). This is true for all batteries, whether it's a large battery pack for an ebike or a simple AA battery for a

> Fun Fact: You could actually run an ebike on AA batteries. It would take over 180 AA's to power an average ebike though you would probably want to consider using rechargeables and buying in bulk!

flashlight.

Using the AA battery as an example, let's talk about specific battery ratings. Each AA battery is 1.5 V and coincidentally also about 1.5 AH. Alternatively, AAA batteries are smaller and while they have the same voltage of 1.5 V, they have a smaller capacity at around just 1 AH. C batteries, on the other hand, are much larger than both AA's and AAA's and while they have the same voltage of 1.5V, they have a much larger capacity of around 6AH to 8AH. This same idea holds true for your ebike battery. Standard ebike battery voltages are 24V, 36V and 48V but are available in different capacities, usually between 8 AH to 20 AH, although larger and smaller capacities are available as well.

The level of voltage and capacity of your battery pack has different affects on your ebike. The voltage is directly related to the speed of your ebike while the capacity is related to how far it will go on a charge. Increasing the voltage of the pack increases the speed of the motor. Increasing the capacity increases the range. The change is roughly linear, with additional losses at higher speeds due to air resistance and other parasitic losses. That is to say, doubling the voltage of the battery nearly doubles the speed of the motor while doubling the capacity nearly doubles the distance your ebike can travel.

Now that we've learned how battery choice can affect your ebike, let's take a look at our options. When ebikes first began gaining in popularity, lead acid batteries were the standard. Nickel cadmium (NiCd) batteries and then nickel metal hydride (NiMH) batteries were introduced in ebike designs in the last decade in an effort to save weight but all three have largely been replaced by lithium batteries. Lead acid batteries are still regularly used in ebikes when cost savings are the determining factor. Unlike the newer lithium battery packs for ebikes, NiCds and NiMHs are hard to find in battery pack units ready-made for ebike conversions.

In sum, this means that your first decision when it comes to a battery will be deciding between low cost yet heavy lead acid batteries and more expensive yet lightweight lithium batteries. Initially, the low-cost of lead-acid batteries is quite attractive, but you will want to keep in mind the downsides. Not only are lead acid batteries much heavier but they also lose their capacity much faster, often in as little as a few hundred recharges. This translates to buying new batteries once a year or more if you use your ebike daily. Also, you need to recharge lead acid batteries immediately after you finish using them to avoid permanently damaging them.

Lithium batteries can be as much as five times more expensive than lead acid batteries but have a useful life of at least five times longer. This means that in the end, you could end up spending the same amount of money for each battery through the course of its effective life, but with lithium you'll have a much smaller, lighter battery. Lithium batteries will also be easier to mount to your bicycle because most come prepackaged with mounting systems specifically designed for electric bicycles.

As you can see, I'm quite partial to lithium batteries. Don't worry if you want to go the lead acid route, I'll cover the necessary information for those as well, but as with all new ebike builders whom I advise, I recommend going the lithium route. There are only three cases where I recommend using lead acid batteries: 1) for a first time builder who is just really unsure of what voltage he or she wants to work with and would like a cheap way to experiment with multiple voltage levels; 2) someone who wants a short term battery to begin with before upgrading to better components down the road; or 3) someone building an electric tricycle where weight and battery storage space are not an issue, but cost is a priority.

Now let's get the info on these lead acid batteries out of the way.

Lead acid batteries

The type of lead acid batteries used in ebikes are called Sealed Lead Acid batteries (SLAs) because they are sealed in a plastic case to keep any acid from leaking out. This means they can be stored in any orientation. Flipping over a non-sealed lead acid battery such as a car battery, on the other hand, would have disastrous consequences.

Standard SLAs come as 12V batteries (or occasionally 6V) that you'll have to wire in series to achieve your final voltage. Series wiring involves connecting the positive terminal of one battery to the negative terminal of the next battery. A wiring diagram can be seen below.

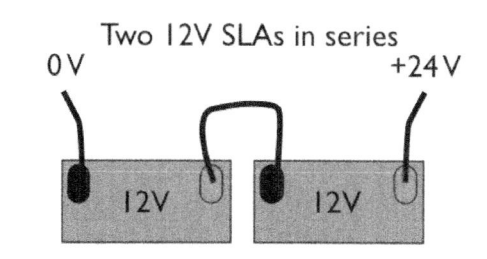

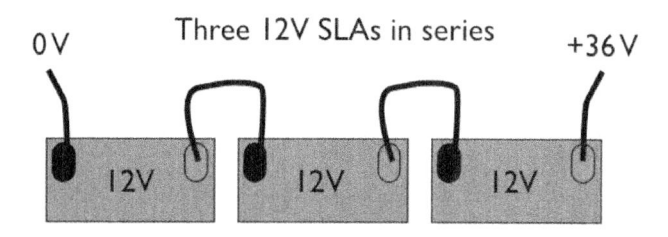

Two 12V SLAs in series
0 V   +24 V

Three 12V SLAs in series
0 V   +36 V

Two batteries in series will make a 24V pack, three batteries in series make a 36V pack, four batteries in series make a 48V pack, and so on. To connect the tabs on your SLAs you can either solder wire directly to the tabs or use quick release connectors. For information on connectors and soldering, see Chapter 6. Make sure to use heavy (thick) gauge wire for your battery pack, at least 16 AWG, though 14 AWG or 12 AWG is even better (thicker gauge wire is actually a smaller number -- it's counterintuitive, just go with it).

When connecting your SLAs together, start with the positive (red) tab on one battery and wire it to the negative (black) tab on the second battery. This will create a 24V pack. If a 24V pack was your end goal, then you're finished! If you need a 36V pack, wire the positive tab on the second battery to the negative tab on a third battery. This will create a 36V pack. Continue wiring your SLAs together like this until you've reached your desired pack voltage.

To check that your wiring is correct, use a multimeter or voltmeter to check the pack voltage by placing the black test probe on the negative pad from the first battery and the red test probe on the positive pad from the last battery in your pack. The total voltage of the pack as indicated on the voltmeter should be the sum of the individual 12V packs.

Choosing the right AH (amp hours) for your SLAs will depend on how far you want to ride your ebike without charging. Most ebike batteries range from 10 to 20 AH. 10 AH is good for shorter trips around town while 20AH is better for longer joy rides or commutes. The nice thing about SLA's is that they are cheap enough to start with a smaller pack and upgrade later if you should decide that you want more range.

One important bit of information for SLAs is that to get the longest life possible out of the batteries, it is best not to discharge them more than 50% depth of discharge (DOD). More than 50% DOD can harm the battery's health and significantly reduce its useful working life. A good rule of thumb is to assume that if you want a usable 10AH of capacity, buy 20AH of battery to be safe and treat the battery well by not discharging more than 50% DOD on a regular basis.

Also, you'll want to make sure you recharge SLAs immediately after you've finished using them. Leaving them drained can permanently degrade their energy storage capacity. One hour here or there isn't going to kill them, but consistently leaving them drained for many hours or overnight will degrade the batteries much faster than normal and will result in a shorter battery life than specified by the manufacturer.

SLAs are very, very heavy (thanks to the 'L' part of SLA). You'll want to make sure that you plan on a strong and secure method of attaching them to your bike. Some people have had success building custom battery boxes for SLAs that mount onto the frame, but most people simply mount them in panniers, which are bags that mount on either side of the rear rack of a bicycle. You can place them on the actual rack as well, but by putting them in the panniers you help lower the center of gravity of all that weight and thereby improve the handling characteristics of the bike.

I've seen many cheap racks break under the load of heavy SLAs, so it is important to make sure you use a good, strong rack that can take the stresses of the heavy batteries bouncing around while riding.

As a side note, you'll also want to invest in a good kick stand, potentially a double kick stand like those used by scooters and mopeds. SLAs are heavy enough that the standard kickstand on most bikes simply can't support the load. You don't want your heavy ebike falling over and hurting someone or damaging your components.

Unlike lithium batteries meant for ebikes, SLAs usually won't come with a charger when you order them, so you'll want to pick up an SLA charger as well.

They come in standard 12V increments so just choose the appropriate voltage charger for whatever pack you plan to build.

If possible, try to find a charger that comes with a matching connector that you can connect to the battery pack. If the charger doesn't come with a matching connector then you may need to search for the connector or buy two matching connectors, one for the battery and one to swap onto your charger. The connector for charging the battery should be wired as

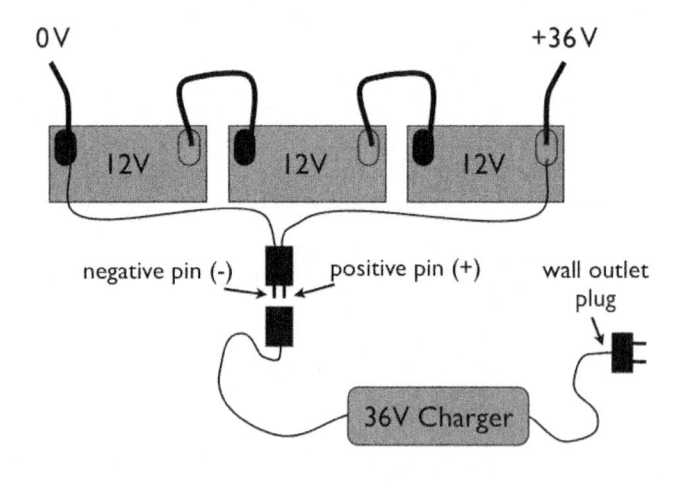

shown in the diagram below.

To check that the connector is installed correctly, use a voltmeter to measure the voltage from one pin of the charging connector on the battery to the other pin. The voltage should be the sum of the 12V batteries in the pack (or slightly higher - a fully charged 12V battery is usually closer to 13V). You'll also want to make sure that you've correctly matched the positive and negative pins on the battery's charging connector to the charger's connector. Use a voltmeter to ensure that you have wired the polarities correctly. If you plug in the charger and the indicator light does not signal that the battery is charging, you have likely installed the charging connector backwards. Check it and if necessary, switch the polarity.

Lithium batteries

Now, let's talk lithium. Lithium batteries require a good deal less work because you don't need to build your own custom pack. Lithium batteries come preassembled and ready to install on your ebike. You also have many more options to select from when choosing lithium batteries.

There are many lithium battery packs that are specifically designed for ebikes and therefore are easily mounted on a bicycle frame or bicycle rack. The first major decision you'll need to make is which version of lithium chemistry you want to use. Currently, the two main options are Li-ion (usually LiNiMnCoO2) or LiFePO4.

Now let's delve into the details of these two lithium battery chemistries. Li-ion batteries are less expensive, slightly smaller and lighter, but are rated for fewer charging cycles, usually 500-800 cycles. LiFePO4 are more expensive, slightly larger, but typically rated for at least 1,500-2,000 charge cycles, meaning one battery could last you five years, even if you use it nearly every day. LiFePO4 is also the safest of the lithium batteries. In the event of a catastrophic overcharging or puncture, it can't oxidize fast enough to cause a fire or create an

A few types of ebike lithium batteries

explosion risk like most lithium batteries.

The next decision you'll need to make is the style of lithium battery you'd like to use: shrink wrap lithium batteries or hard case lithium batteries.

Shrink wrap batteries are exactly as the name implies: the lithium battery cells are enclosed in a plastic shrink wrap shell to protect them from the elements. Hard case batteries are actually the same shrink wrapped batteries, but the shrink wrapped battery is further enclosed in a plastic or aluminum case to give the battery additional structural support and provide an attachment system to the bicycle, as well giving the battery a more finished appearance. Hard case batteries also typically incorporate a locking mechanism to deter theft.

As you might guess, the plain shrink wrapped batteries are typically less expensive than hard case batteries, but require you to get creative with a mounting solution. Many people just use a triangular bicycle bag, but others choose to build a custom battery box to hold the shrink wrapped battery pack on their ebike.

There are various styles of hard case batteries, each with their own mounting mechanism. The most common style works by having the hard case slide onto a metal or plastic guide bolted the bicycle's rack or frame. This allows the battery to be easily removed for charging or swapping with a freshly charged battery.

A variant of this battery type is the "bottle style" battery, which mounts to the existing water bottle bolts on the bike frame, although this requires you to remove the water bottle holder. The "bottle style" battery can also be mounted elsewhere on the bicycle without use of the water bottle bolts, but you'll have to get a bit creative.

ebike with frog battery

Another popular type is the "frog style" battery, but this battery is only appropriate for folding bicycles which have tall, exposed seat posts. The "frog style" battery uses a mount that is secured to the seat post directly under the bicycle seat and allows the battery to slide on and off the mount. An advantage of this style of battery is that it still permits the bicycle to fold but leaves the rack free to use for other purposes.

Nearly any lithium battery specifically designed for use with ebikes will come with a matching charger. Generally, the supplied charger will be a low power version. You can check the approximate power level of the charger by the amp rating which will printed somewhere on the unit. A charger with an output of

1-2A (amps) is a slow charger while chargers rated at 3-5A are faster and more powerful.

To calculate the estimated charging time based on the specs of the charger, use the AH information discussed above in this section. If you have a 10AH battery and a 1A charger, it would take about 10 hours to fully charge the battery (10AH divided by 1A = 10 hours). If you use a 2.5A charger in the same scenario, it will take about 4 hours for a full charge (10AH divided by 2.5A = 4 hours).

There is one more lithium battery technology to discuss. This last group includes lithium polymer batteries (called 'lipos') typically used in remote controlled (RC) aircraft. These RC lipos have the advantage of being much smaller, lighter, less expensive and more powerful than the other lithium chemistries, however, they come with a serious downside: RC lipos are notoriously dangerous! While accidental overcharging or physical damage of LiCoMn or LiFePO4 batteries could mean you've destroyed the pack, a similar error with lipos usually ends in a fire, explosion, or both. Don't believe me? Do a quick youtube search using the terms "lipo", "puncture", "overcharge", and "fire".

For enthusiasts who know what they are doing, RC lipos can successfully be used with ebikes to provide additional power with less weight. But their safe use demands an in-depth understanding of these batteries and is outside the scope of this ebook. RC lipos are great batteries when used properly, but the chance of a dangerous outcome is simply too great for beginners. If you want to explore RC lipo batteries for yourself, you can find links in the resource section of this chapter, but I do NOT advise novice ebike builders to go this route. Your ebike experience will be far less fun and rewarding if you have succeeded in burning down your garage.

The last and final option for lithium batteries is to build your own battery pack. This is not a beginner project, but it isn't that difficult either. Lithium battery cells are commonly available, and they can be combined in series and parallel just like the SLA bricks we talked about earlier in this section. This sometimes requires more complicated tools though and the process is not as straightforward. There is too much detail regarding custom lithium battery packs to include in a single chapter. I wrote an entire how-to book on the subject. It's called "DIY Lithium Batteries: How To Build Your Own Battery Packs". If you are seriously interested in building a custom battery, I'd recommend checking out the book to learn more on the subject.

Controller

Simply put, the controller is the brains of your ebike. Without it, your ebike doesn't know how to operate. The motor is dumb, all it knows how to do is spin. The battery is similarly dumb, all it knows how to do is provide power. But the controller, connected in between the battery and motor, works all of the magic to apply the proper amount of power from the battery at the proper time and feed it to the motor. The

Pick a controller, any controller

controller is also the component that ultimately determines how much power your ebike will have, because the controller is the gatekeeper that actually delivers the power from the battery to the motor. This is a very important concept to understand. Even with a 1,000W motor, if you connect it to a 250W controller, all you have is a 250W ebike. The watt rating on your motor doesn't matter in determining the final power output of the completed ebike; it is the controller that determines the power of the whole system. If you come from the gas engine world, imagine a big 500 hp engine that you're feeding with a tiny little carburetor or injectors with a narrow manifold. If you choke it back like that, you'll never get your full 500 hp. The same thing happens if you use a low power controller on a high power motor.

To be clear, it is not necessarily bad or wrong to connect a low power controller to a higher power motor, it just means that you aren't using the motor's full potential - it's just a waste. Some people actually do this intentionally though because with this configuration the motor is less likely to overheat as there is less power flowing through it than it is capable of handling Basically, it is over-spec'd.

However, going in the other direction can be a problem. If you connect a 250W motor to a 1000W controller, you will have built a 1000W ebike. But, because the motor was only designed to handle a quarter of the power being supplied to the motor, it is highly likely that the motor will burn out quickly when placed under load. Exactly how quickly depends on many factors including the quality of the motor, weight of the rider and the type of riding being done. This particular scenario could see the motor being destroyed in anywhere from a few hours or days to a few weeks or months.

Now that you've seen that the power rating of the controller ultimately determines the power of the ebike, you need to learn how to evaluate the actual power level of the controller. Controllers are often marketed as having a specific watt rating. It is easy to find controllers online that are specified as 350W, 500W, 800W, etc. However, this is typically not the actual peak watt rating of that controller, but rather the average watt rating. Power is calculated using the simple formula where Power = Volts x Amps.

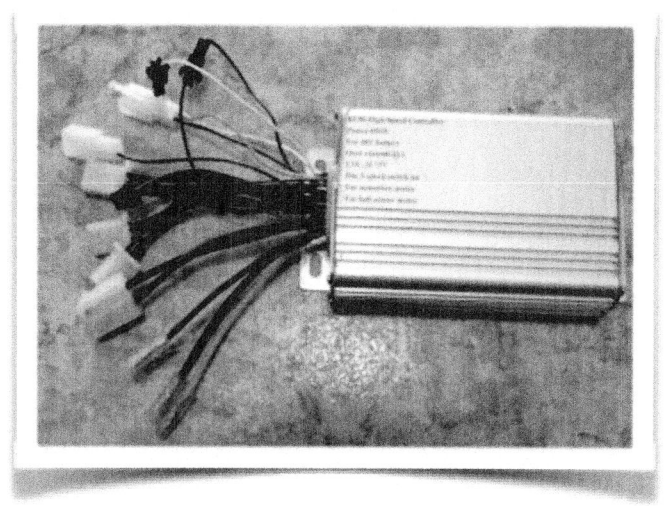

"450 Watt" ebike controller

Let's take a look at this sample controller. This controller is marked as a 450W controller, but if we look at the actual specifications, we see that its peak current is 30A.

This means that when peak current is being supplied (which happens when accelerating from a stop or climbing steep hills) this controller is actually capable of delivering instantaneous power equal to 48V x 30A, or approximately 1,440W. It is very important to calculate the actual power of your controller to make sure that you aren't pairing it with a motor that is insufficient to handle that power. For example, if you intended to use a 350 watt motor on your ebike, you might see this controller and say "450W isn't much greater than 350W, I'll use this controller and get a bit more power". But in reality, 1,440W is much greater than 350W and the extra power that you'll be getting will last for about 90 seconds before your motor is completely destroyed. This is why it is crucial to make sure that you calculate the actual power level of your controller. Burning up your brand new motor is a really bad feeling!

When searching for controllers online, you are stepping into a bit of an abyss. Almost all ebike controllers come in a nearly identical aluminum box, so telling them apart can be very tricky. High quality controllers rated to take thousands of watts look basically the same as cheaper controllers that should stay below 1,000 watts. While price isn't the best indicator, you can use it as a rough approximation of quality. A high quality controller, built with premium components meant for pushing the limits of power, will cost upwards of $100, while cheaper controllers can cost as little as $25.

If you are planning to build a low power ebike, i.e. something less than 1,000W, then you don't really need to invest in a top of the line controller because pretty much any controller that can be purchased off eBay or other sites online will work okay. For top of the line controllers, custom manufacturers, like Lyen, produce great quality controllers that can really hold up to some abuse. For first time ebike builders, a $50 controller will serve you well.

Matching the controller to your ebike specs is an important step. One parameter that sometimes causes problems for first time ebike builders is the controller's Low Voltage Cutoff (LVC). The LVC is a built-in setting found in most controllers that allows it to shut down if the battery voltage dips below a predetermined level. As you ride your ebike the battery voltage decreases as it drains. The LVC keeps your battery from being drained too far, a potentially damaging condition for the battery.

LVC's vary from one controller to another, so you'll need to identify the specific controller specs before purchasing it. If you have a 36V battery and want to use a 48V controller, your battery voltage generally won't be high enough to power the controller above the LVC. Therefore, it is essential to buy a controller that is matched to your battery.

Some controllers are marketed as multi-voltage, and allow you to connect batteries of different voltages. A common multi-voltage controller is a 36V/48V controller. This type of controller may have an extra wire that when connected will set the controller to 36V and when left unconnected sets the controller to 48V. Using a multi-voltage controller gives you the advantage of being able to change your battery later without the need to buy a second controller. But make sure that you follow the instructions that come with a multi-voltage controller in order to select the proper voltage to match your battery.

Throttle

The throttle for your ebike functions essentially like the gas pedal in your car, the more you push it (or turn it, in the case of an ebike throttle), the faster you go. Not all ebike throttles are created equally. There are three main form factors for ebike throttles: full twist; half twist; and thumb throttles.

Full twist throttles are what you think of as standard equipment on a motorcycle or scooter, the whole grip twists in your hand when you turn the throttle. The full twist throttle gives a definite motorcycle feel that many people enjoy, but others find that it can be tiring on the wrist to hold the throttle for long periods of time. Also, full twist throttles can be dangerous if the ebike is left on while the user is pushing the bike because the grip can accidentally be twisted and cause the bike to launch forwards. This instance is rare, but usually happens at the worst times, such as when you are walking your bike down a set of stairs or through a doorway.

Half twist throttles are made in two pieces. The first piece is the throttle grip and the second is a simple rubber bicycle grip. Both are about 3 inches long. The two pieces fit together so that when you twist the throttle, you are only twisting the half that is closer to the center line of the bike (half twist, get it?) instead of the entire grip. This provides that same "motorcycle throttle feeling" as the full twist, but it means that when you are at your preferred throttle level, you can grab onto the fixed rubber half of the throttle to hold the throttle at that set point. This can save your wrist from getting tired. The half twist is also a bit safer when

pushing your bike around your garage or

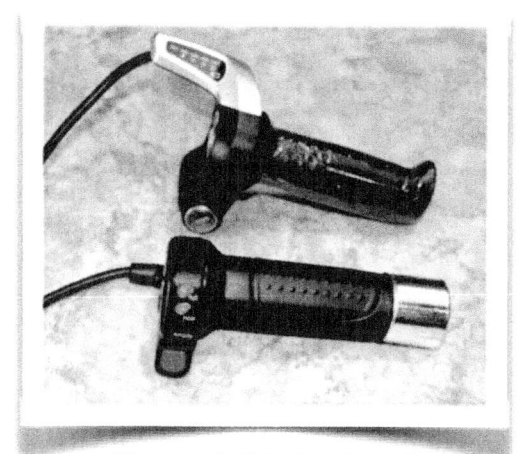

Two different full twist throttles with LED battery gauges. Top throttle has keyswitch, bottom throttle has button

apartment, because you are less likely to accidentally twist the half throttle if you've forgotten to turn your ebike off at the switch.

Half twist throttles are my favorite, because when I'm cruising at full speed I can hold onto both the half twist section and the half of the rubber grip next to the throttle. This means I am using the palm of my hand to hold the throttle in the 'wide open' position without straining my wrist as with a full twist throttle.

Both half twist and full twist throttles take up the same amount of space at the end of the handlebar, taking into consideration both portions of the half twist. If you have a shifter on the right hand side of

Half twist throttle with matching grips

your handlebars, you will either need to move it further up the handlebar and accept that it will be a little less comfortable to operate (although you'll likely be using it less) or switch it to the other side of the handlebar where the shifter may work in the opposite direction because it may be upside down. If you have a shifter on each side, you may want to look at a thumb throttle.

Thumb throttles differ from half and full twist throttles in that only a small lever extends up for your thumb to press. Thumb throttles are small and allow the original bike grips to be used. Also they create less interference with shifters and other handlebar accessories. The downside is that your thumb needs to remain on the lever and many people complain of getting a sore thumb after holding the thumb throttle down for long periods of time.

Thumb throttle with matching grips

One additional interesting disadvantage to thumb throttles only applies to people living in cold climates. In use, your thumb will be hanging down below the handlebar and exposed to the wind. As a result, your thumb will get colder much quicker than the rest of your hands. One of my ebikes in Pittsburgh had a thumb throttle and I found that in weather below freezing, my thumb would feel like an ice cube in about 10 minutes, even through gloves!

Throttles are also available with a range of accessories including LED battery indicators, buttons and key switches. The battery indicator is a nice way to know how much of the energy of your battery pack you've used, although because it functions simply by taking a voltage reading, it is not incredibly accurate. I always tell my ebike students to assume that this type of battery indicator is only accurate to within about

+/-15% of the pack charge. So if it is showing 50% pack level, you can assume that you have an actual level of between 35%-65%. If you think that seems like quite a wide range, then you're correct. Using the voltage to measure battery pack capacity is simply an approximation. Commercial electric vehicles determine pack capacity by using complicated mathematical algorithms combined with sensors measuring everything from current draw to battery pack temperature. A $10 ebike throttle just isn't that sophisticated. Instead of thinking of it as an exact battery meter, consider it as providing a general indication of your battery charge. However, after a period of getting used to your ebike, you'll be pleased to see just how well these LEDs, combined with your own feel and knowledge of your ebike's performance will help you understand just how much range you have left.

    A throttle with a button or key-switch is a good way to utilize the enable/disable wire of your controller (which we'll discuss more in Chapter 4), if your controller is equipped with this feature. The button or key-switch can be wired so that when pressed or turned, the controller turns on, and when depressed or returned, the controller turns off. This is handy if you use a battery pack that doesn't come with its own case and thus does not have an on/off switch or button. The key switch obviously provides more security than the button type, but button throttles are more common and less expensive.

If you don't have the enable/disable feature on your controller, you can also wire the switch inline with the ground wire on your throttle (the black wire). This will not turn off the controller, which means it will still be drawing a small amount of current, but it will keep the throttle from functioning when the button is pressed or the switch is turned. Keep in mind that the more accessories your throttle comes with, the more wires it will have and the more complicated it can be to install. But don't let this scare you off. Accessories help you customize the form and function of your ebike.

In addition to throttles, there is another way to control the speed of an ebike, the Pedal Assist System (PAS). PASs are more common in Europe due to laws requiring their use. The PAS mounts on the bicycle's crank, which requires removal of the crank, a process which typically involves special tools. The PAS works by sending a throttle signal to the controller when it senses that the user is pedaling. In order to make the ebike begin to power itself, you have to begin to pedal, which the controller senses through the PAS and then responds by sending power to the motor. In my opinion, the PAS is annoying to install and even more annoying to use. It's the equivalent of a car with a foot pump instead of a gas pedal -- in order to keep the car running you'd have to continually pump the foot pump instead of just holding a gas pedal down. Who would want that? Well, for those do who have their heart set on using it either for the exercise aspect or if it is required in by law in your region, I'll cover PAS's in some detail in Chapter 6.

Accessories

There are a number of other ebike accessories that can make your ebike even more fun and useful.

Lights

First, you may want to consider adding a special ebike light. This isn't a standard bicycle light running off AA batteries, it is a stronger light that uses your ebike's battery for power. There are several different types, from simple lights that mount to your bike and are hardwired to the battery (which means they run all the time, even during the day) to lights that come with switches, electric horns and even battery meters.

If you choose to use a simple throttle without an LED battery meter, I would suggest that a dedicated ebike headlight unit with its own power switch and battery meter might be a good choice for you. A good ebike headlight is made by Wuxing and is widely available from various online suppliers. I usually buy mine from BMSbattery, although you'll see a list of different vendors in Chapter 4.

My favorite headlight is a CREE XML T6 LED headlamp/headlight, meant for off-road bikes. This one is actually a knock-off of the more expensive MagicShine MJ-808E, yet costs only $25. The light is powered by an external lithium battery pack that is small and easy to hide somewhere on your bicycle frame. I also recommend getting a wide-angle filter lens for the headlight to spread the beam from a concentrated spot to a wide bar of light that better illuminates the road ahead. From Amazon, the filter is only another $5, meaning you can get an entire, super bright wide angle headlight for just under $30 total. This is a great deal. Another bonus is that it comes with a headlamp style attachment meaning if you ever go hiking or camping and want to be able to give squirrels or other small animals permanent vision damage, you're all set.

If you decide to not go with a large headlight on your ebike, some form of front and rear light should be considered a must. Even a $5 bicycle tail light can save your life. For a greater level of safety, I also use colored lights that fit between my spokes. I like them because I enjoy the circles of light they make when my wheels are spinning and because they make it MUCH easier for people to see me at night. I use two sets of lights for each wheel. One light usually looks like a light spinning in a circle but two lights installed across from each other give the illusion of a complete circle of light when the wheel gets up to speed. At 20 mph or more, it looks quite like a Tron cycle!

Mirrors

Mirrors are a good addition to an ebike if you plan on driving on the road along with cars. Not that you'll be doing a great deal of merging, but it's still nice to know when you have a car coming up on you from behind.

Most bicycle mirrors are generally poor quality and their use on ebikes that travel at speeds greater than normal bikes only compounds the problem. You'll want to look for a good strong mirror that doesn't have issues with vibration when traveling at high speeds. The kinds with bendable necks often vibrate too much. Search instead for a mirror with a strong and rigid mount to the handlebar. My favorite is the Mirrycle mountain bike mirror.

Storage

Some type of storage on your ebike can be helpful for carrying your charger or stashing a few important items. Many people find their ebike is so convenient that they end up using it to replace their car for small errands and light-duty grocery shopping, but this requires some storage space.

A front bag is a nice addition to the handlebars. It doesn't have that '12-year-old-girl' look of a front basket and still allows you to stash a few things up front. A rear rack and panniers (bags or boxes that hang on either side of a rear rack) are excellent for grocery shopping. You can easily fit 4 or 5 bags of groceries with this set up.

A triangle frame bag fits in the (you guessed it) triangular part of the frame. It is a great way to add storage space without increasing the width of your bike like panniers, plus the weight is in the center of the bike, front to back, and relatively low in terms of center of gravity. This makes for better handling.

For some customers that wanted more secure storage, I've installed motorcycle and scooter trunks on a rear rack on their ebikes. These provide a lot of storage space as well as offer a great location for lights and reflectors, but can make it harder to swing your leg over the bike, especially if your frame already has a large stand over height.

Seat

A comfortable seat is a very important factor in the overall the comfort of your bike and its ride quality. Because you'll probably be doing less pedaling than normal bikes, consider switching over to a seat designed for a cruiser bike. These are larger and more comfortable than standard bicycle seats.

While you're at it, if your bike doesn't have suspension, you might consider adding a suspension seat post. Make sure you find the correct size for your bike as there are a number of different standard sizes for seat posts. The size (diameter) should be listed on your seat post near the bottom of the post. It's measured in millimeters, and will look like "28.2" or similar. Just slide your seat post all the way out to check the size, then search for a suspension seat post of the same size. Remember, 'be kind to your behind.' You'll be glad that you did.

Tires

Your ebike will likely be going faster than a regular bike and covering much greater distances, so upgrading the tires will be a quality investment. Additionally, strong tires are less likely to get flats, which is an additional pain to fix on an electric bike because removing the hubmotor wheel is not as quick and easy as removing a regular wheel. Basically, better quality tires are a good idea and will save you a great deal of frustration down the road. If you're working on a budget, then consider just upgrading the tire on the hubmotor wheel.

The type of tires you choose depend on the type of environment in which you live and the type of driving you plan to do. For people intending to ride their ebike off-road and on trails, knobby tires will give you the grip that your motor needs for good handling. For people commuting on the road or sidewalk, street tires are more appropriate. Knobbies on pavement can create slight vibrations that vary in intensity between types of tires.

Additionally, you'll want to make sure you're getting some quality rubber. If you are buying your tires in person at a bike shop, feel the rubber between your fingers and flex it. You want something that feels strong yet is still pliable. Hard or brittle rubber is a sign of a cheap and/or old tire and should be avoided.

If you purchase your tires online, make sure you choose a reputable brand and read the reviews. Any well known, quality brand is going to use good rubber. It's only when you're considering something on the bargain basement end, like a no-name imported Chinese tire, that you'll need to make sure the reviews don't indicate poor quality rubber that will fail quickly, and often at the worst possible moment. Though there's rarely a good time for a tire to fail.

Of the hundreds of tires I've changed on hundreds of ebikes, I have seen one company whose products rise above all the rest: Maxxis. These tires are great quality with thick rubber that lasts for a long time and resists punctures well. There are a number of different Maxxis tires for different conditions. My personal favorite is the Maxxis Hookworm, a great street tire that is smooth enough for excellent grip and traction with a good tread pattern for rainy day wet roads. If you're looking for an off road tire then Maxxis has you covered there too with a number of different models, but I love the Hookworms for their all around street use.

You may also want to invest in other anti-flat tire measures such as thicker inner tubes and anti-flat inner tube sealant. Inner tube sealant is squirted into the inner tube through the inflation valve and sloshes around inside. When something punctures the tube, the sealant fills the hole. It works well most of the time, but

> Quick Tip: If you are unlucky enough to have something small puncture your tire like a staple, thorn, etc - AND you have some anti-flat gunk in your tube - just leave the foreign object in place. It will help seal the hole as the gunk flows in around it. I've had a thumb tack in my rear tire for the last two months just hanging out, but I have Joe's Tire Sealant in it so I'm fine!

it still isn't going to stop every single hole, just most of them. Even so, I highly recommend either Green Slime or Joe's Tire Sealant. Consider it some inexpensive peace of mind.

While I'm on the subject of avoiding flat tires, here's another tip to think about while you're driving. While many people prefer to ride their ebike in the middle of the lane, you would actually do better if you try to stay on the part of the lane where the car tires track. That means avoiding the center section of the lane as well as the extreme edges. Most of the debris that gets scattered on the road, such as metal scraps, thorns, nails and other nasty refuse, gets knocked around by the automobile tires until it comes to land in the areas where the tires don't contact the road as much, i.e. the center and edges of the lane.

If you look carefully next time you're out, you'll often see the road surface is darker in the center and on the extreme edges, with two cleaner wear patches on either side of the center of the lane - exactly where car

Drive on either side of the center of the lane

tires roll along. You can use this as a guide for where to locate your bike on the road. I like to use the patch on the right side of the center of the lane because it allows me to move over easier when a car approaches, and makes it easier for cars to pass me if I don't hear them coming up from behind me.

Lock

A strong lock is a must for an ebike. You've spent a lot of time and energy, not to mention money, building your new ebike and you don't want someone else reaping the benefits of your resources.

The first rule of bike locks is that there is no such thing as a theft-proof lock. If someone wants to steal your bike bad enough, they will. That's an important concept to remember. Any time you lock your bike up outside, you're taking on some risk. The whole idea of using a bike lock is to make it as difficult and frustrating as possible for a potential thief to successfully steal your bike. This means that the thief would either need to spend a lot of time working on the lock, use a tool that makes their actions obvious to passersby, or both.

Cable locks are immediately out of the question. They can be cut through in mere seconds. During the time I worked in electric bike shops, customers would come in from time to time asking us to cut off their cable locks after they lost their keys. They were amazed to see me open the lock in 3 seconds with a pair of bolt cutters. Here, see for yourself how easy it is: http://youtu.be/P32x3RqsNmY?t=4s.

As a result, there are two main options remaining: a U-lock or a chain lock. Both locks require an angle grinder or oxyacetylene torch to remove (with the exception of the cheapest U-locks and chain locks which can also be opened with very large bolt cutters). I've removed both U-locks and chain locks for customers using an angle grinder, but it takes at least a minute or two and basically sounds like someone is repeatedly banging a rock against a piece of metal at an insanely fast speed, which is essentially what is happening. Anyone in a 100 meter radius would easily hear someone cutting through one of these locks.

That's the point. If it draws a lot of attention, a thief is less likely to attempt it.

The best U-locks are made by Kryptonite, with the strength of the lock increasing from the grey, to the orange, to the yellow locks. My all time favorite U-lock is the Kryptonite 18mm New York Fahgettaboudit U-Lock. It has a massive shackle and a great locking mechanism. The only problem is that it's a bit stubby as compared the standard 10-inch long U-locks, meaning your tire has to be right up against the fixed object if you're locking the wheel. The 10 inch yellow Kryptonite U-lock allows you to be a little further from whatever you're locking on to, which translates to a little more real world versatility.

There are a number of companies that produce good chain locks, just look for a thick one. Kryptonite is also a good source for quality chain locks but because these are a simpler lock, pretty much any thick chain and strong lock will work about as well.

Another thing to remember is to make sure you're locking the right part of your bike. How many times have you seen a single bicycle wheel locked to a street post or fence? That's usually because someone locked their bike by just the wheel which was held on by a quick release axle. The thief made off with the bike in literally seconds by simply pulling the quick release lever on the wheel axle - no tools required! When possible, always try to lock part of the frame of your bike, and never lock just the wheel if you have a quick release axle. Even if you have bolt-on axles, it only takes a moment to loosen the axle nuts and away goes your bike minus one wheel.

When I lock my bike, I always try to think like a thief. I ask myself, how would I steal this bike. If I lock part of the frame, that means I'd have to actually cut through the bike frame to steal it, making the bike fairly worthless (though I could still steal it just to strip it for the parts - like I said, think like a thief!).

But don't just think like a thief, think like an experienced thief! You don't necessarily have to cut through the bike OR the lock to steal the bike. It's so frustrating when I see people lock their bike to a chain link fence or some other flimsy structure. Any dimwit with a pair of handheld wire cutters could be through a link in that chain link fence in seconds and take your bike home, where he has all the time in the world to work on the getting the lock off the bicycle frame in his garage. Make sure whatever you are locking to is a stronger material than that of your bike frame. Steel sign posts, fence posts, parking meters and hand rails are usually good options. But in the case of an isolated signpost, before you lock to it, try pulling it up first. Urban bike thieves routinely loosen the sign post from the base to create a sneaky trap. One quick pull up on the signpost and your ebike is their ebike. Think like a thief!

The single best anti-theft tip I can give you is this: even better than using one strong lock is using two strong locks. Remember, the goal is to show a thief that it just isn't worth their time to try and steal your bike. Maybe (and that is a big 'maybe') they can get through one lock without being noticed, but if they see two locks, chances are they aren't going to take the risk of investing the time to defeat the first lock if they'll have to do it all over again without a guarantee of success for the second lock. I'll admit, even I don't use a second lock sometimes because it can be annoying to lug around, but the times that I know I'll be locking my bike in a "bad part of town," I always make sure to bring along the second lock. Even if you just compliment a U-lock with a small cable lock, the second lock will add a certain amount of deterrent value to any would-be bicycle thief.

Of course the best method is just to park your bike indoors whenever you can, though for many trips this simply isn't possible. In these cases, two locks are your best bet to make sure your investment remains yours.

Tools

The last thing you want is to be stuck on the side of the road with a flat tire or other bicycle problem and no way to even attempt a fix. If you already have storage on your bike, consider throwing in a ziplock bag with a few screwdrivers, a wrench that fits your wheel nuts, a spare inner tube and any other parts you think you may need.

If you don't have storage on your bike, consider an under-the-seat bag, also called a saddle bag. These are very small and could be perfect for holding a few tools. Murphy's law certainly applies here; if you bring the tools, you probably won't need to use them.

I've known a few people who take a water bottle, cut the top off, fill it with tools, then tape the top back on and put it in the water bottle holder. This sneaky tool holder insures that you'll always have what you need when you're away from home, but keeps prying eyes and fingers off your stuff as well.

Speedometer

A speedometer isn't absolutely necessary, but it certainly is a fun addition to an ebike! There are analog speedometers meant for regular bikes that mount on the front wheel with a little tab that spins in the spokes, but these are rarely strong enough to hold up to the speeds of an ebike. In my experience, most of these analog speedometers meet the trash bin in a matter of weeks or months of ebike use.

Digital speedometers are more appropriate for our level of use and abuse. Choose one with a good sized screen with big numbers. The last thing you want when you're flying down a hill is to take your eyes off the road for too long while trying to determine just how fast your speedometer says you're going. Bragging rights are important, but remember you can always check your "max speed" screen later!

Another helpful feature is checking your ride distance. This can assist you in determining your ebike's range and help you compare different riding styles and conditions to see how they affect your range. I'll give you a hint: pedaling helps your battery last longer.

Horn/bell/warning device

If you plan on riding in an urban area surrounded by pedestrians, at the very least, you'll want a strong bicycle bell. If you'll be riding out on the road though, your bell will be as helpful as a whisper in a windstorm. There are electronic horns designed for regular bikes, which usually use AA batteries, but most of these aren't that much better than bells when it comes to the decibel department. If you want a really loud electric horn, get a moped or motorcycle horn. Most are 12V, which means you'd need a DC-DC converter to connect it to your ebike battery, but you can find many 36V and 48V versions. I use a 48V moped horn on my bike and it's perfect.

There is still one bicycle horn left that sets itself apart from the rest: the AirZound. This air horn is by far the loudest horn I've ever seen for an ebike. It's actually so loud that I don't recommend using it as a "you there on the sidewalk, get out of the way" horn. It's really more of a "you there in the car, don't run me over" horn. I recommend it to anyone spending any amount of time riding in traffic. I've used one many times to keep a car from pulling out of a driveway or a parking lot entrance into my path. Drivers are looking for other cars and aren't expecting a bike, that they may or may not see, to be traveling so fast towards them.

Kickstand

You'll want a good strong kickstand for a heavy ebike. Depending on where you mount your batteries, you may be able to get away with using your stock bike kickstand if it is strong enough. Most standard bicycle kickstands that come with cheap bicycles are too flimsy. Don't trust them if your ebike winds up being overly heavy or if you have your batteries mounted fairly high on the bike.

If you need a stronger kickstand then I recommend a double kick stand. These are the style of kickstand used on most mopeds and motorcycles. It allows your bike to stand upright instead of leaning to one side and is better for soft surfaces like grass where a single kickstand can sink in and cause the ebike to fall over. Think about it: if it's good enough for a motorcycle, it's good enough for an ebike too!

Of course, if you're on a tight budget, you can get away with a single kickstand, it's just a bit less stable. You can reduce the severity of the instability problem by mounting your batteries as low as possible on the bike. This makes for a lower center of gravity that will also help with the stability of your ebike on the move.

Air pump

It's not a bad idea to keep a handheld air pump on your bike. Most pumps have a mount for strapping to a bike's frame. The front fork is a good location for a mini pump because it keeps the pump out of the way but gives you the option of topping off your tires while you're on the road.

You'll want to keep those tires full to help with efficiency and protect against getting a flat.

> Quick Tip: I'd recommend getting a decent floor pump to keep at home as well. Always keeping your tires filled up means less chances for flat tires, which are an even bigger pain in the rear end for ebikes.

While you're at it, try to keep a schedule of checking and inflating your tires every two weeks or so. Inner tubes are strong, but they are also microporous, meaning they slowly leak air naturally over time.

Alarm

There are ebike specific alarms available that come with a handheld remote, just like a car's alarm. They work on vibration and are powered by your ebike's battery. There are also a few stand-alone models that use a 9V battery and a code instead of a remote.

For most people this is overkill, but if you find yourself regularly parking your bike outside for long periods of time, an alarm can add a bit of peace of mind and supplement a good lock. I wouldn't recommend using only an alarm though. Rather, pair it with a stout lock for extra protection.

Watt meter

There are various types of watt meters on the market that are appropriate for use with ebikes. A watt meter allows you to monitor the power of your ebike, as well as a few other important parameters such as battery voltage and the remaining capacity of your battery.

There is one top of the line watt meter built specifically for ebikes called the CycleAnalyst, or sometimes just "CA" for short. The CycleAnalyst not only allows you to see all the parameters of your battery and motor usage, but also includes an integrated speedometer. The unit is built into a backlit case for easy night viewing and is designed for simple mounting right on the handlebars.

Other watt meters have similar functions to the CA but don't have speedometers. Also they aren't designed to be easily mounted on the handlebars, though this can still be accomplished with the creative use of some zip ties.

Helmet

I'm not trying to be your mother, but I am going to be blunt here: if you ride an ebike, it is a very, VERY good idea to wear a helmet. Most states have bicycle helmet laws of some kind. The ones that don't, well, they should.

Just about every study ever performed has shown that helmets make a significant difference in the injury rates and survivability of bicycle accidents. At the increased rate of speed you'll be traveling on your ebike, not wearing a helmet is pretty much just stupid. If that offends anyone, think of it this way: it's better to be offended and still alive than the other way around.

When I started riding ebikes, I began with a full face motorcycle helmet. I did a lot of off-road riding (and was probably more cavalier than I should have been) and simple things like wet grass or loose gravel were enough to make me think "man, I'm sure glad I have this full-face helmet" as my body was sliding across the ground. A bicycle helmet can keep your head from cracking open but it won't keep your face from being scraped off by the cheese-grater-like road surface.

During my time living in the Middle East, where the

This guy might look like a nerd, but he's a very much ALIVE nerd!

temperature
can get quite toasty outside, I would often forgo the full face helmet for a standard bicycle helmet, though I am certainly aware of the risks and rewards of each option and I know I'm making an informed decision with full knowledge of the risks involved. When I choose to put on the standard bicycle helmet though, I often make a

mental note to not push my luck when slicing and dicing through traffic columns that day.

Ultimately, as with every option on your ebike, the helmet issue is up to you. But hey, if your spouse asks, I can at least say that I strongly advised the use of a helmet.

Endless Sphere

I'd like to introduce to you to a great ebike building resource, Endless Sphere, or ES for short. ES is an internet forum specifically for electric vehicles with a main focus on electric bicycles. It can be found at www.Endless-Sphere.com/forums. It is simply the single largest meeting place for ebike enthusiasts anywhere in the world or on the internet.

Everyday, hundreds of people contribute to the wealth of knowledge already stored on ES. Whenever I have a question or problem that I can't quite resolve, I head to ES and do a search to see if anyone else previously has had the same issue. Most of the time, the answer is yes, and I can see exactly how other people have helped to solve the problem.

Occasionally a search won't turn up the answer, and I'll start a new topic. Within hours I usually have responses from great ebike builders from all over the world who are quite willing to help me sort out whatever I'm working on. All of this info is saved on the forum for future reference by anyone with a similar question. Inevitably, the end of threads accumulate posts along the lines of "Just had this same problem and now I've found the solution here, thanks!" or something to that effect. I can't recommend ES enough as a resource to ebike builders, both at the beginner and advanced levels.

ES is by far the single best resource I can provide to you to help with any of your unanswered questions. The people are incredibly knowledgeable and helpful and the response time is phenomenal. I highly recommend taking a look at the forum and try searching around for topics that have interested or puzzled you about ebikes.

> Quick Tip: Remember to do a search for your question before just posting it up on Endless Sphere. There is so much information on the site already that your question has probably already been answered.

Ebikes.ca Hubmotor Simulator

One final set of resources that I want to offer to you (at least before you get to the formal resources section) is the ebikes.ca Hubmotor and Ebike Simulator, available at: http://www.ebikes.ca/simulator/. This website is an invaluable resource for testing simulations of different combinations of ebike components without having to buy all the parts and do all the work. The values generated by these simulations are derived from the actual manual tests that Grin Technologies has done on all these hubmotors so that you don't have to! But the best part of all is the price: it's totally free! The guys at Grin Technologies love ebikes so much that they put these kinds of resources out there for all of us to help grow and expand the community and its collective knowledge base.

To use the simulator, simply select your components from the drop down menus on the left side of the screen, including motor, battery and controller. You can also create a custom battery or controller if you don't see an option with your exact specifications among the list. Then make sure you select your correct wheel size for your bicycle (26" for most standard bicycles, 20" for most folding bikes) and approximate the weight of you plus your ebike, together. Lastly, choose the units of measure you'd like to use and click 'simulate'.

You'll be given a chart with a series of colored lines. These lines represent torque, power, load and efficiency. While these are interesting figures, what will be more useful for you is the maximum speed and maximum range. The speed is displayed at the bottom of the chart, directly below the vertical dotted line. The range is shown in a box at the bottom right side of the screen.

I've spent hours playing around on the simulator, virtually testing different designs before I build them. This is a great way to choose the right motor and battery combination, or see how changing a controller from 20A to 30A can affect your power or acceleration curves.

Chapter 3 Resources

- Adjusting V-brakes: http://youtu.be/mGgidUE8drE
- Adjusting disk brakes: http://youtu.be/XAUTCZ3gIyU?t=17s
- Mid-drive motor info: http://www.electricbike.com/mid-drive/
- Mid-drive source and installation video: http://california-ebike.com/
- Friction-drive motor info: https://sites.google.com/site/commuterbooster/
- Another friction drive option: http://www.eboo.st
- My favorite bicycle lock: http://www.amazon.com/Kryptonite-997986-Black-Fahgettaboudit-U-Lock/dp/B000OZ9VLU
- SUPER loud air horn: http://www.amazon.com/Delta-Airzound-Bike-Horn/dp/B000ACAMJC
- CREE XML T6 Bicycle Headlight: http://www.amazon.com/gp/product/B006Y1FK18
- Wide angle filter lens for headlight: http://www.amazon.com/Angle-MagicShine-Gemini-Lights-Headlight/dp/B004WLCLQY

- Lock: http://www.amazon.com/Kryptonite-997986-Black-Fahgettaboudit-U-Lock/dp/B000OZ9VLU

- Cycle Analyst: http://www.ebikes.ca/drainbrain.shtml

- RC Lipo resources: http://www.endless-sphere.com/forums/viewtopic.php?f=14&t=19956

- Ebikes.ca Hubmotor Simulator: http://www.ebikes.ca/simulator/

- Endless Sphere electric bicycle forum: http://endless-sphere.com/forums/index.php

Chapter 4: Purchasing Components

Now that you've planned out your ebike based on your individual specifications, its time to locate and purchase the parts. There are two basic ways to go about this: buying a kit with all the parts included or buying the individual parts separately. For first time builders, a kit is usually the best way to go. By buying a kit, you'll receive most or all of the parts that you need to build your ebike, except for the bicycle itself. Also, with a kit, you know that the parts will be compatible with each other.

Kits are usually sold in two styles: with a battery or without a battery. If you buy the kit with a battery, it will likely have matching connectors for the battery and controller, making it even easier to install. If you buy a kit without a battery, and source the battery separately, you may need to install your own connectors on the battery if the stock connectors do not match your controller.

Even though buying a kit with a battery included is usually easier in terms of installation, there are a few reasons why you may want to purchase your battery separately. For one, some kits are sold with lead acid batteries. As we discussed in Chapter 3, lead acid batteries are not the best option for ebikes. Also, even if the kit you are looking for comes with a lithium battery, you may be looking for a battery with higher capacity or you may have found one at a better price elsewhere. Therefore, buying a kit without a battery can sometimes be advantageous.

When searching for kits online, you'll see the kits usually referred to with a certain power rating, such as a "500 watt kit". This watt rating is essentially a controller watt rating, and the kit will pair the controller that is appropriate for that wattage motor. Just because a kit is called a "500 watt kit", doesn't mean it actually puts out 500 watts. Likely, that's an average rating for the controller, and the kit has a maximum instantaneous power of closer to 750-1000 watts. As discussed in Chapter 3, knowing the true power of the controller in the kit will help you choose the right kit for you.

Even with the supplier's standard power labeling scheme, you can compare kits based on one supplier's wattage labels. For instance, if a website sells a 350 watt kit, a 500 watt kit and a 1,000 watt kit, even if the actual power of the kits varies slightly from the name, you can bet the 500 watt kit is more powerful than the 350 watt kit and less powerful than the 1,000 watt kit.

If you are comparing the kits from one site to another, it can get tricky because one seller's "500 watt kit" could be equivalent to another seller's 350W or 600W kit. If the seller doesn't list the specifics of the controller, namely the current rating, ask them. If they can't or won't tell you this information, walk away. This is an indication that they either don't understand the parts well enough to provide knowledgeable advice or are selling inferior parts. Either way, there are many other reputable sellers from which to purchase.

Double vs single wall rim. Note the exposed nipple heads in the single wall rim

You'll also want to make sure that the kit comes with a quality laced wheel. Make sure the seller uses "double wall aluminum rims". Double wall rims have two layers of metal instead of one, meaning the spoke nipple heads are sunk under the top layer and give the added bonus of posing less of a puncture hazard to your inner tube. Single wall rims are used on many cheap pedal bicycles and are not strong enough for ebike use. They'll work, but they often fail before the first 1,000 miles. Do yourself a favor and start with a good double walled rim for your hubmotor wheel. Single walled rims are fine on the non-motor wheel.

Also, check to make sure the wheels use stainless steel spokes. 14g is the minimum thickness for quality ebike spokes but 13g is better and the strongest motors (greater than 1,000-1,500W) sometimes use 12g spokes (the smaller the gauge number, the thicker the spoke). Check with the vendor to verify the specific materials used in the wheel, if that information is not already listed. One thing to note here though is that there are many different kinds of stainless steel spokes and they vary greatly in quality. One of my favorites are Sapim spokes but most asian companies that sell hubmotors will use a lower quality stainless steel. This usually won't be a problem unless you live in an area that is tough on corrosion and/or you have an unusually strong motor. The road salt in Pittsburgh combined with my 2,000 watt motor led to a few corroded and broken stainless steel spokes after only a year of use on one of my bikes.

Another important point to keep in mind: make sure that the kit comes with a cassette or freewheel (set of bicycle gears) already on the motor if you're getting a rear hubmotor. Some kits supply just the motor without a freewheel and this requires you to visit your local bike shop to pick up the appropriate freewheel to add to the motor.

There are many websites that sell ebike kits, and I'm going to mention a few here that I have personally used. I'll list other options in the resource section of this chapter, but the ones I'm listing here are ones that I can personally review based on my experience working with them and their components.

EM3EV

EM3ev.com has a number of different ebike kits including both hubmotors and mid drives. They carry the popular MAC geared motors and also have the Bafang mid-drive kits as well. In fact, the new version of the Bafang BBS02 comes with upgraded components based on the recommendation of Paul, the owner of EM3EV. So no matter where you get your BBS02 from, you can thank him for helping Bafang improve their product.

Perhaps the product that EM3EV is best known for is their batteries. They carry a wide range of sizes and shapes, all made with only high quality Samsung battery cells. Other batteries are often a mystery, even those sourced from the US, because it's hard to know what battery cells are under the cover. But Paul only uses the highest quality cells and his packs are well known in the industry for their longevity and quality.

One unique feature about EM3EV is that it is located in China but run by a British ex-pat. By being located closer to the source of many ebike parts and manufacturers, EM3EV can offer lower prices than western companies while still providing western levels of service and support. A real win-win for price and quality. If you want to learn more of the backstory of this enigma of an ebike company, check out this article: http://www.ebikeschool.com/inside-look-at-ebike-parts-supplier-em3ev-com-and-its-founder-paul/

ECityPower (AKA: BMSBattery.com)

BMSBattery.com has a number of kits available for sale, from 250W up to 1,000W. Because they sell many different ebike parts, you'll want to navigate to the pages with the complete kits, which are found in the list of categories on the left hand side of their webpage. Click "Bike conversion kit" and then you will be able to select from the subcategories "Front-Driving" and "Rear-Driving", and then further down select between the sub-subcategories of 24V, 36V and 48V.

Almost all of the kits on BMSBattery.com are geared hubmotor kits. BMSBattery sells Cute motor kits which are referred to on their site as Q85, Q100 and Q128 kits, as well as Bafang kits which are referred to as SWXU, K5, and QSWX kits, and the powerful Bafang BPM kits. These kits vary from 250-500 watts, with the exception of the Bafang BPM kits which can be run up to 1,000 actual watts when using a 48V battery and controller. They also sell the Q11 kit, which is listed as 1000 watts, but at 48V with the supplied 30A controller actually peaks at 1,500W.

Most of my experience with BMSBattery is with their Q series kits. The Q85 and Q100 kits are great quality, though I and others have had problems with the Q128 motor, which has seen a decrease in quality over the past few years.

The Q11 is a great motor kit if you want a powerful ebike, but may be more than most people are looking for in their first bike. The Q11 is basically identical to a Nine Continent (9C) motor, sharing the same parts but without the 9C logo on the case. In the ebike world, the exact origin of some parts can be a bit of a mystery, but these Q11 motors are likely made either in the same factory as the 9C's or elsewhere using identical 9C parts.

BMSBattery.com has good prices compared to other sites, but their shipping tends to be very high. For this reason, I would recommend picking up a few inexpensive accessories, such as spare throttles, controllers, lights, etc, because they won't add much to your shipping and are dirt cheap. Other places sell throttles for $15 or $20; BMSbattery.com sells the same throttles for a fifth of that price.

When ordering a kit from BMSBattery.com, you'll need to type the wheel rim size in the box at the bottom of the page and click 'save' in order to add the kit to your cart. They offer a variety of throttles so you can put a note in your order form with a specific request for your preference of throttle type. In my experience, if you don't specify the type, they usually ship with a full twist throttle by default.

While I have generally been very happy with kits from BMSBattery.com, the one complaint I would make is that their wheels are sometimes not expertly built. Occasionally I receive wheels with spokes needing to be tightened and/or requiring some minor truing on arrival. This is true of most Chinese suppliers because quality control standards simply don't compare to those in the western world. If you pick up a kit from BMSbattery.com, check your spokes when your kit arrives to make sure they are all tight, and then check them again occasionally throughout the life of your bike (though you should be doing this anyway; more on this in the maintenance section). If you ever hear a click-click noise from the wheel, it's likely a loose spoke telling you to perform a check.

If you should find a loose spoke, you can either tighten it yourself by turning the nipple (the nut at the end of the spoke that hangs out of the rim) with a spoke wrench or gently with a pair of pliers if you don't have a spoke wrench. BMSbattery.com sells spoke wrenches for a couple of bucks and I highly recommend picking one up just in case. If you ever have a number of spokes loose at once, or you notice your wheel doesn't look quite straight when you spin it, perhaps with a side to side movement on each turn, consider taking your bike to a bicycle shop and asking them to true the wheel. You can do this yourself with a little practice and a quick youtube video search, but if you don't feel comfortable taking on this task yourself, ask a professional for help. Loose spokes are a runaway condition - the longer you leave them the quicker more will loosen. Each loose spoke puts more stress on the others until finally they can't take it anymore and pop goes a spoke, followed soon by more. Preventative maintenance is the key here.

Grin Technologies (AKA: www.Ebikes.ca)

Ebikes.ca sells a number of kits from mid-range to high power kits. Their geared hubmotors are the model eZee motors, which have received quite positive reviews. These kits are some of the more expensive geared motor kits available, but the quality control standards are excellent (they ship from Canada, not China).

Another advantage of ebikes.ca is that the kits ship with an inner tube and tire on the wheel already, so you don't even have to swap your own tire on. This makes them that much easier to install, though you'll probably want to learn to remove your own tire anyway in case you ever need to replace a flat.

Ebikes.ca also sells Nine Continent and Crystalyte direct drive hubmotor kits. These are more powerful kits and perfect for those of you wanting to build a sporty, high powered ebike. Again, these kits are more expensive than other good alternatives, but the service and quality control at ebikes.ca is second to none. And if you want a geared kit, they also have a few options from Bafang and TDCM. Lastly, they've developed a really neat all-axle hubmotor in house that is worth checking out!

Their founder, Justin, is practically a celebrity in the world of ebikes. Justin and Grin Technology's other employees are quick to answer personal questions via email and provide expert advice. It's kind of like in the early days of Apple when you sent a tech question in and got a letter back with the answer to your problem from Steve Wozniak.

There is no better customer service in the ebike world than from Grin. Justin also invented the Cycle Analyst display mentioned in Chapter 3. That awesome ebike watt meter (and battery gauge, and speedometer, and so much more...), while available from multiple vendors, is available for the best price from ebikes.ca - straight from the source!

Grin also has tons of extra parts. Whether you need electrical connectors, tools, bike parts, basically almost anything for an ebike, they've got it. They also make some of the best torque arms on the market, going through several revisions over the last few years, so I'd definitely recommend checking out their torque arms if you are looking at a motor of over 500 watts or using an aluminum bicycle frame. Torque arms are something we'll talk about more in Chapter 5. For a quick preview though: torque arms help hold strong motors in place so they don't spin inside your bike's dropouts.

I almost always recommend ebikes.ca to first time ebike builders due to their superior parts, customer service, and commitment to the ebike community.

Ebikekit.com

Ebikekit.com has a smaller selection of parts meant for electric bicycles and tricycles. They are one of the few companies still pushing lead acid batteries, while the rest of the industry has moved on to the modern standard of lithium batteries. Ebikekit.com does offer standard lithium batteries with their company logo applied to the case, but at prices often three or four times the price of an equivalent overseas lithium battery.

Ebikekit.com used to be the only US sources for the popular MAC geared motor, mentioned earlier in the motor section of Chapter 2. These MAC motors are a great choice for someone who wants the advantages of smaller and lighter geared motors without the lower power limit that plagues most geared motors. However, others such as LunaCycle now offer the same motor and for nearly half the price. Some newcomers to the ebike hobby have also reported that Ebikekit.com's support is abrasive and less helpful than others.

Yescomusa.com

Yescomusa.com sells ebike kits that are no-brand motors but generally have proven to be of good quality. In comparison, their rims and spokes leave much to be desired.

When I receive a motor from Yescomusa I usually ditch the single wall rim and lesser quality spokes and replace them with better quality components. For a low power ebike they are probably fine, but for a high power bike, they simply won't hold up very long.

I've also found that the controllers from Yescomusa are subpar in quality. Again, for a low power ebike, they could be acceptable, but for a high power ebike, you could find yourself with a failed controller if you push your ebike to the upper end of its limits.

Goldenmotor.com

Goldenmotor.com is the maker of the MagicPie and SmartPie ebike conversion kits, arguably some of the simplest to install conversion kits available on the market today. When run at 48V, the MagicPie is also one of the more powerful stock kits you can put on your bike, peaking at around 1,500 watts of instantaneous power. The beauty of the MagicPie kit is that the controller is actually built into the motor and all of the connectors are designed to come together into one connection block. Simple!

The kit is so easy to install that I'd definitely recommend it to anyone interested in a simple, powerful ebike conversion kit

Golden Motor offers a number of different battery options that are compatible with their kits. Search around their site and you'll see all the accessories they offer.

ElectricRider.com

ElectricRider.com is a great source for the powerful Crystalyte motors. These are motors that are pretty much as powerful as you'll ever want to go for your first ebike build. Bigger and stronger motors exist, but I do not recommend them for the novice. My very first daily driver ebike kit came from this site. It was a Phoenix Cruiser kit, which at the time was a Crystalyte 5304 motor, famed for its crazy high torque level and resistance to heat. Unfortunately Crystalyte discontinued this line of motors a few years ago and replaced it with a new line of motors that weighed half as much but with a slight sacrifice in power and torque. Even so, these are still great, powerful motors for anyone looking for a high power setup.

These new kits remain really great quality. Another advantage is that you'll be getting a well laced wheel, one of the most important considerations when comparing kits. I've also been highly impressed with Electric Rider's customer service. They've always been quick to answer my questions, a real plus, especially for the first-time ebike builder.

Another big plus about Electric Rider is that their website fully explains precisely how fast you can expect each kit to go, based on the different wheel sizes, and they provide real world data to back up their claims. They told me my Phoenix would do about 29 mph and, sure enough, it got exactly 29 mph. Other less reputable dealers might try to round up to that magic 30 mph number figuring, "what does one little mile per hour matter?" Well, it matters a lot when it represents the honesty of the company you're working with, and that's why I've always felt entirely comfortable ordering motors from Electric Rider. It's a company I know that I can trust. Their prices can be a bit higher than other vendors, but I consider that a good tradeoff for better motors, parts, and service. You get what you pay for.

Most of my experience with Electric Rider has been with their motors; however, I've heard good things about their lithium batteries. I know they waited a long time to even offer a lithium battery. But now they found and sell one that they guarantee is capable of meeting the demands of their powerful motors. It's that commitment to quality and customer service that ranks Electric Rider up high in my list of best suppliers.

Kinaye Motorsports

Kinaye is a relatively new US-based ebike component and kit company, but has already made quite a name for itself. Opening in early 2015, they focus namely on higher power components and kits designed for high performance ebikes. They are one of the only US sources for the popular 3,000W MXUS direct drive motors found on many high power ebikes. In addition to powerful motors, Kinaye also supplies the high performance controllers, batteries and other components needed for such demanding applications. The company has also recently been introducing other ebike components and accessories including high-end bicycle parts and frames, making for a truly one-stop shop. While a lot of these parts might be more than the average first time ebike builder is looking for, make sure to check back when you feel the need for speed, as Kinaye has everything needed to get you laying down rubber quickly.

LunaCycle

LunaCycle is fairly new on the scene but is one of the most affordable USA vendors. They sell a number of parts ranging from motors and controllers to accessories, but one of their most visible products are their homemade batteries. They use high quality lithium battery cells to produce a number of different sizes and shapes of battery packs. They've also got a neat charger that can do different levels of charging to help preserve the lifetime of a lithium battery. I haven't actually ordered anything from LunaCycle yet, so I can't attest to their support and ordering process, but there are a lot of other reviews out there by people that have done a lot of business with them already. Just google "LunaCycle Review" and you'll find many.

California Ebike

If you are looking for a BBS02 mid-drive kit (and that's the one I recommend if you want a mid-drive) then you should check out http://california-ebike.com/

for a great US source. Owned by Doug, California Ebike has developed a reputation in the industry for fair prices, excellent support and innovative design. Doug developed the first popular conversion process for adapting the BBS02 to fat bicycles, a common build request that was impossible to do without tricky custom alterations until Doug developed his product. It was subsequently ripped off and now available from other vendors, but the best quality can still be found from California Ebike. They offer quite competitive pricing as well. Of course no one can compete with the drop-shipping prices of China, but if you want a local source that comes with warranty and support for only a small premium over Chinese supplier's prices, California Ebike is the place to find it.

Other vendors

There are a number of other vendors out there selling ebike kits. A quick google search will yield dozens of sites. I simply haven't had the opportunity to try them all.

Additionally, kits of all types, performance levels and qualities can be found readily on eBay. But here I offer a caution: I've seen some great motors come from eBay sellers and I've also seen some complete junk too. Without having previously purchased a kit from a particular seller, it is impossible to know exactly what you'll be receiving. You can check a seller's feedback and reviews to get an idea of previous customers' levels of satisfaction, but eBay ebike kits remain a "buyer beware" situation.

Assembling your own set of components

If you haven't been able to find the perfect kit with all of the parts you're looking for, buying the parts individually will allow you to customize your ebike to your exact specifications. All of the complete ebike kit sellers that I listed in the previous section also sell individual parts which you can select à la carte for your ebike conversion. Remember, the four essential parts are: a motor; battery; controller; and throttle. Everything else is just an accessory.

Motor

Unless you already know how to lace a bicycle wheel (a rare skill), or you have access to someone that does, or you're prepared to pay for the service, you'll want to buy your motor already laced into a rim. A local bicycle shop can lace a hubmotor into a wheel, but they usually don't treat the wires coming out of the axle with the respect and care they deserve. This can result in broken wires. Few cities currently have an ebike shop with someone who really knows how to open motors and repair damaged phase/hall wires. If you consider yourself technically inclined, lacing your own motor might be something you want to tackle on your own, but it isn't for beginners. For anyone lacing a hub motor, be extra careful with the wires leaving the motor so as to not damage them. The few times I've given a hubmotor to a local bicycle shop to lace for me, I've taped over the end of the axle with the wires to prevent them from being damaged.

Remember the lessons we talked about in the motor section of Chapter 3; you're looking for a motor in the right power range for your bike along with the right characteristics. Generally, this means a geared motor for a lower power bike and a direct drive motor for a high powered bike, with about 500W-750W being the line of demarcation between low and high power classifications.

Small geared motors are cheaper to purchase and ship, which may influence your choice for your first ebike. Just like for buying a kit though, check with the vendor to ensure that their product incorporates double wall aluminum rims and quality stainless steel spokes. Also verify the speed of the motor at the voltage you plan to use.

Battery

By now you should have decided on whether you're going with SLA's or lithium. SLAs can be bought locally at many outlets. Hardware and electronics stores often carry SLAs. Just make sure that you are getting

12V batteries and that the capacities match up. You don't want to combine two 10AH batteries with a 7AH. They should all have the same AH rating.

SLAs are also available online but be aware that the heavy weight can increase the shipping costs. Usually, lithium batteries for ebikes can only be bought online, unless you live in a city with an ebike shop. However, you should be wary of buying from a local specialty ebike shop. I've worked in local ebike shops and the prices on parts are usually much higher than if you were to buy them yourself online, even including shipping costs. In fact, most local vendors are typically buying parts from the same sites I've listed here, then marking them up and selling them to their customers. Cut out the middle man if you can.

That isn't to say that there aren't honest local ebike shops out there. You should just always do your homework and research the fair market price for batteries before believing whatever the brick-and-mortar store salesman says.

Always compare before you buy. If you live in a country with strict customs restrictions or VAT (value added tax) requirements, buying locally may be your best bet or only option. If you live in the US, definitely shop around online to compare prices. The thing I probably missed the most when I lived outside the US was the ability to have just about anything in the world shipped to my doorstep in a matter of days, no questions asked (or VAT added).

When buying your battery, make sure that you have a plan to mount it to the bike. If it's a lithium battery in

Three different aluminum case batteries

an aluminum case, then it will come with some form of mounting hardware to attach it to either the bicycle frame or the rack. Be sure that you know how it mounts to know if it will fit your bike. Not all batteries fit equally well on all bikes. Do your research.

Rack batteries are generally universal, but anything designed to fit inside the frame can be tricky on smaller frames or bicycles with non-standard geometry (cruisers, full suspension bikes, etc). If you plan to put the battery in a frame bag, make sure you know it will fit. Get the measurements of the battery and the bag from the manufacturers, and leave a little wiggle room just in case. It's kind of like a pair of pants, sometimes the battery might fit but the zipper isn't going to close.

Also check out what charger is offered with the battery. A low power charger, 1 or 2 amps, is actually healthier for the battery than a more powerful 3-5A charger, but will take longer to charge. You may consider getting two chargers if you want the option to keep one at home and one at work, or some other creative approach.

Also, make sure your battery is rated to provide the amount of power that your controller and motor will require. If a battery states that its maximum continuous current rating is 10A, and that it can provide a peak of 20A for short bursts, then this battery is not appropriate for a controller rated at 30A. That controller will pull more amps than the battery is rated for. This will either repeatedly blow the fuse in the battery, which is the best case scenario, or slowly destroy the battery, which approximates the worst case scenario.

Sometimes, battery discharge rates are given in units of 'C', for example 2C. In this case, the 'C' represents the capacity of the pack in amp hours (AH). If the battery is an 8AH pack that is rated for 0.5C charge and 2C discharge, the highest charging current it can take is 4A and the highest discharge current to the controller it can provide is 16A. Similarly, a 20AH pack with the same rating of 0.5C charge and 2C discharge could charge at 10A and discharge at 40A.

This demonstrates that a higher capacity pack can provide more current safely to the controller and motor without damaging the battery. This means that if you have a powerful motor and controller, you may want to look for a battery with a higher capacity than if you were building a lower power ebike.

All of the companies listed in the kit section also sell batteries, but I'll add another couple of battery suppliers here. Ping Battery (www.PingBattery.com) builds and sells quality LiFePO4 battery packs in a variety of voltages and capacities. Ping's standard packs are bricks but Ping can also build custom packs to your specifications. In my experience, Ping batteries display excellent quality and use better components than other battery vendors. Ping Battery prices are also a bit higher than the others. This is another clear example of "you get what you pay for".

A special note on lithium batteries: most shipping services and locations consider lithium batteries to fall within the category of 'hazardous materials' and therefore apply shipping restrictions. These restrictions vary from location to location but can often include extra fees, forms and labeling requirements. Ultimately, this may not effect you at all, but be prepared if the post office or shipping service presses the issue.

Controller

As discussed in Chapter 3, ebike controllers are difficult to judge in terms of quality without knowing precisely who manufactured it. Buying a name brand controller like a Crystalyte, Infineon or Lyen controller is a good way to go if you want to ensure that you are getting a top quality controller that can take years of hard use. Compared to your standard, no-name silver controller, these branded controllers are known to contain higher quality components, but can cost three to five times as much as the no-names.

If you decide to go with a simple controller and you don't plan to push the limits of power on your ebike, a no-name controller is usually an acceptable choice. These controllers are made with components rated to handle less power than the brand name controllers. But like I said, if you want to stay in the 250-750W range, these controllers are usually satisfactory. In fact, you can sometimes find non-branded controllers inexpensively enough that it's actually worth just purchasing two at once in case you have a problem down the road. Also, with shipping from China being so high, you can usually buy four or five controllers for the same shipping cost as just one.

If you are sourcing your parts separately rather than buying a kit, you'll want to check out what kind of connectors you'll have on your controller. Odds are that with an à la carte method of choosing parts, the connectors won't all match your controller. This will still work fine, but it means you'll have to either replace the connectors or just cut them off and solder the wires together yourself. Neither of these things are difficult to do, they just require a bit more effort. Don't let this aspect of the build keep you from choosing components that you really want.

Throttle

The throttle is a straightforward component and you have many choices to customize the functionality of your ebike. Nearly any ebike throttle available on the market will work for you, but you'll want to choose one with the features that are most appropriate for your needs, comfort and pleasure. The simplest is a 3 wire throttle, which has no extra buttons, switches or LEDs; just twist and go! I actually tend to prefer these simple throttles for most applications because with fewer connections there is simply less to break or go wrong.

The next possible upgrade would be a throttle with LEDs to let you know (approximately) how much battery you have left. I say "approximately" because they are very inaccurate. When 2 out of 4 LEDs are lit, it doesn't necessarily mean you've used half your pack. Instead, interpret it to mean that you've used anywhere from 35-65% of your pack, depending on the specifics of the throttle and the internal circuit design that the manufacturer used to measure voltage. These devices are basically good for letting you know whether your battery is nearly full, about half discharged, or nearly dead. Still, having even that basic information can be helpful at times. Nearly empty means it's time to ease back on the throttle and make those remaining few electrons count!

The next step up in throttles would be a unit with buttons or switches that you can use to control features on your controller. The main function is an on/off switch or button. This would be wired to connect the large red power cable and the smaller red enable/disable wire. A button can also be used to turn on lights, set cruise control or set a speed limit, but all of these features will only work if your controller is already equipped with these functions.

Some throttles come with a momentary contact button instead of a push-it-once-for-on, push-it-again-for-off type of button. The momentary contact button only allows electricity to flow while your finger is actually holding the button down. This feature isn't good for lights, speed limiters or other applications where you wouldn't want to have to hold the button down all the time.

I like to use a momentary contact button for regenerative braking,

> Quick Tip: Throttles with green buttons are usually momentary contact buttons. Throttles with red buttons are usually push-once-for-on, again-for-off style buttons.

because I can push the button to slow down and when I release it, the regenerative braking stops. I can also use a momentary contact switch for ebike horns or electric bells. Perfect!

Other parts of the kit

While there are only four required components to make your ebike work: motor, battery, controller and throttle, there are many other accessories that are often included in kits.

E-brake levers are installed in place of your normal brake levers. They work just like your original mechanical bicycle brake levers, but also have a switch inside to determine when you are pulling the brakes. If you connect your e-brake levers to your controller, when you pull the brakes your controller will automatically cut power to the throttle. This prevents you from accidentally trying to give throttle while pulling the brakes, something that you may be more likely to do in a sudden, emergency, panic braking situation.

Also, if there was ever a problem with your ebike that caused you to lose throttle response and pin the throttle to the maximum, pulling the brake lever on the ebrakes should cut power to the motor. This is a very rare and unlikely problem to experience, but if it ever happens to you, immediately come to a stop, turn off the ebike and disconnect the battery. Don't start looking for the problem until you have pedaled home and don't try to fix it on the road.

Ebike headlights often come in ebike kits and offer both a headlight and a battery gauge in one unit. Sometimes a horn is even integrated into the unit as well. Usually the horn will require you to purchase a separate button to engage it, although as I mentioned previously, this can be another good use of the momentary contact button on the throttle.

Some kits come with pedal assist systems (PAS), discussed in Chapter 2. I rarely install these because they are just a pain in the behind, but if pedaling to control your motor is something you are really interested in doing, you can pick up one of these units on the cheap.

The most important consideration when assembling your own custom 'kit' out of individual parts is simply to make sure that all of the components are compatible. This generally means making certain that your motor and controller are of comparable power levels and that your battery is rated to provide the amount of power that your motor and controller require. If the battery vendor says their battery can supply a maximum of 20 amps (which would be 2C for a 10AH battery), using that battery with a 40 amp controller is not a good idea. While it can be done, it will severely limit the useful life of your battery. Instead of a 1,000 cycles, you may be lucky to get 500.

Chapter 4 Resources

Vendors for ebike parts:

- www.Ebikes.Ca
- www.BMSBattery.com
- www.GoldenMotor.com
- www.GreenBikeKit.com
- www.Lyen.com
- www.Conhismotor.com

- www.emissions-free.com/store/
- www.ampedbikes.com/store.html
- www.falconev.com
- www.ElectricRider.com
- http://us.itselectric.ca
- www.PingBattery.com
- www.Leafmotor.com
- www.LunaCycle.com

Chapter 5: Installing the Kit

Installation of an electric bicycle kit is a relatively simple process, although as with most new endeavors, your first time will be a little slower going as you move along the learning curve.

Before you begin the installation, make sure you have a clear, clean work area to lay out all of your parts and tools. You may want to lay down a blanket or towel under the bicycle so you don't scuff up the seat or handle bars if you flip the bike over to make working on it a bit easier. Most ebike conversions should take between 2-5 hours, depending on how many parts you'll be installing and if your build requires any custom fabrication outside of simple bolt-on parts. A good plan is to start in the morning on your first build, just in case you run into snags. With any luck you should have plenty of time for your first ride before dinner!

You'll want a basic set of hand tools available. Here are some tools you might need: Phillips and flat head screwdrivers; 8 mm, 10 mm, 15 mm, 19 mm and 21 mm wrenches (or a single adjustable wrench); bicycle tire levers; regular pliers; needle nose pliers; metal file; 2.5 mm, 3 mm, 4 mm, 5 mm, 6 mm and 8 mm allen keys; rubber mallet; small cutter or snips for zip ties (not a hobby knife); wire cutter; a soldering iron with solder and electrical tape. You may need other tools depending on your specific bike and components, but this is a good start. In actuality, you might not need all of these tools. But I generally make sure that I have all of these tools on hand in case the need arises.

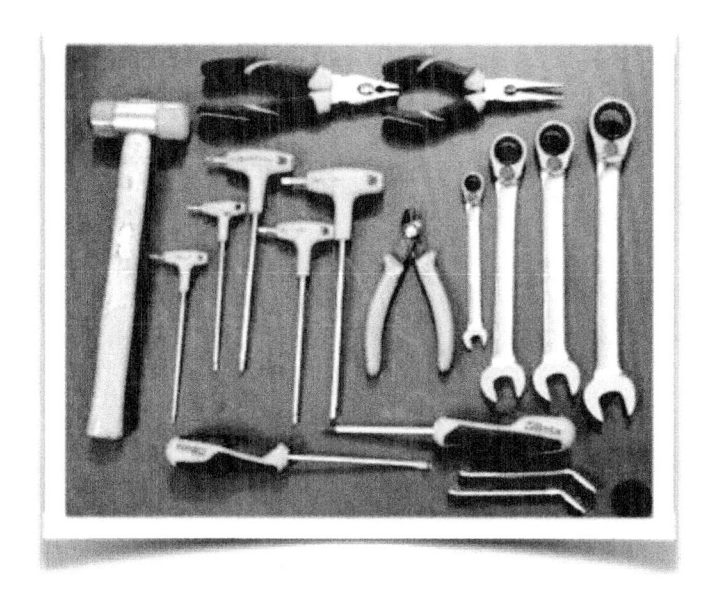

It is also a good idea to plan out your installation before you start. Make sure that everything will fit on the bike in the planned location. Pay special attention to ensure that the intended location on your bike frame of your primary components, particularly the battery and controller, are acceptable in terms of the length of the electrical cable runs. You don't want to get everything installed only to discover that your battery or throttle cable are six inches too short to reach to your controller. You may need to consider relocating the controller or battery to make it all work. Alternatively, you can add a short length of wire to the components, but this takes more work and it introduces another location for a potential electrical-type failure, so best to avoid it if possible.

Installing the motor

The best place to start the conversion is with the motor. To install a hubmotor, begin by flipping your bike upside down. This is where that towel or blanket comes in handy. Remove the wheel, front or rear, that you intend to replace with the hubmotor. If the wheel uses a quick release system, just open the quick release lever. If the wheel is bolted on, a 15 mm wrench is typically the right size to get the job done.

If you have rim brakes, you'll need to open the brakes to remove the wheel, otherwise the tire won't fit between the brake pads. Just squeeze together the brake pads and look for the spot you can pull the cable loose, freeing the brake arms to spread apart and allowing the tire and wheel to pass through.

Remember that there are significant advantages to using a new, stronger tire on your ebike. However, if you plan on reusing the original tire on your hubmotor, start by removing the tire from the rim using bicycle tire levers. If you don't have bicycle tire levers, a couple of flat head screwdrivers will work, but you'll want to be careful that you don't damage the tire or tube with the sharp edges - old, dull flat head screwdrivers would be better. Deflate the tire by depressing the pin in the tire valve. You can use your tire lever or screwdriver for this.

You will hear the air escape and the tire will quickly deflate.

Place a tire lever into the gap between the tire and the rim, then push down to lift the tire up and away from the rim. Place another tire lever 4-5 inches (10-12 cm) down the tire from the first lever and repeat. This time the section of tire between the two levers should pop free from the rim. Remove the first lever and continue using it further down the tire to pry the tire up and over the rim. Repeat this process a few times, until you can get a finger in the space and remove the rest of the tire from the rim. Work your fingers all the way around the tire until that side pops free of the rim. Now you can pull the other side of the tire off the rim as well. If it's stubborn, use the tire levers again.

Before you can put a tire and tube on your hubmotor's rim, you'll need to cover the holes from the spokes so that the sharp edges don't cut into the inner tube. You can either remove the rubber rim strip from your original wheel and place it on the hubmotor rim, or use electrical tape to go around the inside of the rim a couple times, covering the holes. If your old wheel's rim strip is aged and cracking, electrical tape might be a better solution. If you use the electrical tape though, you'll need to cut a hole for the tire valve to pass through.

A properly installed rim strip

Once you've got the spoke holes in the rim covered with a rim strip or electrical tape, place one side of the tire on the rim. You'll need to stretch it just a bit to get the tire bead over the edge of the rim. Make sure you take note of which direction the tire is supposed to spin. Some tire treads work equally well in either direction, but others are designed to shed water or grip loose dirt and are designed to go only one way to work properly. If the tire is designed to rotate in a certain direction, this will be indicated on the side of the tire, usually with an arrow indicating the proper direction of rotation.

Now, inflate the inner tube slightly, just enough to let it take on a circular cross-sectional shape, not enough for it to actually stretch like a balloon when inflated. This makes the tube less floppy and easier to work with when you try to fit it into the tire. Place the side of the tube with the valve inside the tire and push the valve through the valve hole in the rim. Push the rest of the tube inside the tire and begin pulling the bead of the tire over the rim wall to close the tire on the rim.

If your tire is stubborn, you may have to use the tire levers again to stretch the last bit of the tire over the rim. Make sure you don't pinch the tube or you can cause a flat. If the tube is getting caught between the tire and the rim, you may need to deflate it just a bit.

If your hubmotor came with a bicycle freewheel or cassette (the bicycle gears) installed on the motor, you're already set. If not, you'll need to install them yourself. This is as simple as screwing the freewheel onto the motor shaft (remember: righty tighty - lefty loosey). Go slow to make sure the threads engage correctly and you don't cross thread it. You should feel little to no resistance for at least the first 3 or 4 turns of the gears while screwing them on. If you feel immediate resistance, you've cross threaded it and will need to back it off and try again.

Taking the freewheel back off is more complicated than putting it on. If for some reason you need to take it off, you'll need a special freewheel removal tool. This slides down the center of the freewheel, over the axle, to engage the central part of the freewheel. Using a wrench, you can turn the tool counter-clockwise to unscrew the freewheel. A freewheel that has been on the wheel for months or years may be very difficult to break free. But if you simply cross threaded the freewheel for a turn or two when installing it, it should come back off fairly easily.

If you don't have a freewheel removal tool (most people don't) and can't bring your wheel to a local bicycle shop to have them remove the freewheel, in a pinch you can use a flat head screwdriver against the teeth inside the central part of the freewheel. Just brace the flat head screwdriver against a tooth and gently tap the backside of the screwdriver with a mallet. Keep in mind, this only works if the freewheel was installed

6-speed freewheel and removal tool

'finger tight' and not screwed all the way down with any force. If you are removing a freewheel that has been fully seated and tightened down, there is no substitute for a freewheel removal tool.

If you are replacing the front wheel of your bike with a hubmotor, and the original wheel was held onto the bicycle using a quick release axle, you'll need to prepare the fork to accept the hubmotor. Your quick release fork has a feature called "lawyer lips" which are protrusions around the holes for the axle designed to keep the wheel from falling out even if the quick release opens. The lawyer lips presumably got their name due to their suggested inclusion by manufacturer's lawyers weary of lawsuits caused by wheels falling out of forks due to improperly closed quick release axles.

The problem with lawyer lips is that they only work for quick release axles and will interfere with the nuts on your hubmotor axle from properly closing on the fork. You can either use a "C" washer, available from ebikes.ca, or file/grind off the lawyer lips so they are flush with the bike. This way the nuts on your hubmotor axle can close tightly against the fork and will not be stopped by the lawyer lips. But be careful not to remove too much material or you'll overly weaken the dropouts. Remember, any hubmotor more powerful than 350 watts shouldn't go into the front fork, unless you've got a torque arm on there as well.

"C" washer

If your bicycle has disk brakes, remove the disk from your old bicycle wheel and install it on your hubmotor. If your hubmotor came with bolts already installed for a disk, use those bolts. If not, use the bolts that came with your original bicycle wheel. There should be an arrow on your disk indicating which direction it is meant to turn. Make sure to install it correctly and not reversed.

Now you're ready to install the wheel on the bicycle. Make sure to pay careful attention to the wires coming out the side of the motor. You need to be careful with these wires during the installation because if you damage them, the motor won't work properly and the problem will be very difficult to repair.

I always flip the bike upside down to install the hubmotor. Rest the bike on the seat and handlebars on a blanket or towel. If the bike feels unstable, try raising the seat height to effectively lift the rear end up.

If your hubmotor is designed for the rear wheel, you'll need to lift up the chain and place it over the gears on the motor. Then let the weight of the motor do the work by lowering the motor into place above the drop outs with the bike still upside down. Turn the motor to allow the flats on the axle to line up with the flats of the dropouts. Sometimes it helps to place a 10mm wrench on the flats of the axle and use the wrench to turn the axle to line it up with the dropouts. Again, be careful with the wires.

Be sure that you are placing at least one washer between the motor axle's flat shoulder and the frame. You don't want the frame resting just against the shoulder of the axle. Instead, you want a washer there to provide additional clamping surface.

If the motor's axle doesn't slip down into the dropouts even when you move it back and forth with the 10 mm wrench, you may need to file the dropouts to be slightly wider. Often it is just the thickness of the bicycle paint that can interfere with the fit of the axle. Don't get too carried away with the file. Work it a bit and then check the motor fitment again. Continue until the axle fits snugly down to the bottom of the dropout channel.

It is also a good idea to install the motor with the wires exiting the axle towards the ground (or towards the ceiling when the bike is upside down). This creates a "drip loop" so that when water splashes onto your wires, it drips down from the loop and doesn't run along the wire and down into your motor. Use a cable

> Quick Tip: Make sure you secure your motor wires tight enough with cable ties so that if the bike were to fall over, the wire is secured back behind the end of the axle's hollow hole and not severed by its sharp edges.

tie to secure the wire against the bike frame to form the drip loop.

Depending on the strength of your motor and the material that your bike frame is made from (aluminum vs. steel), you may need to install one or two torque arms onto your motor axle to help resist the motor from spinning in the dropouts. A strong motor in a soft aluminum frame can actually spread the dropouts apart, allowing the motor to spin in place, twisting and even cutting the wires exiting the motor. This situation is more common for a front motor than a rear because the fork is weaker, but can still easily happen in the rear dropouts if the motor is strong and the dropouts comparatively weak.

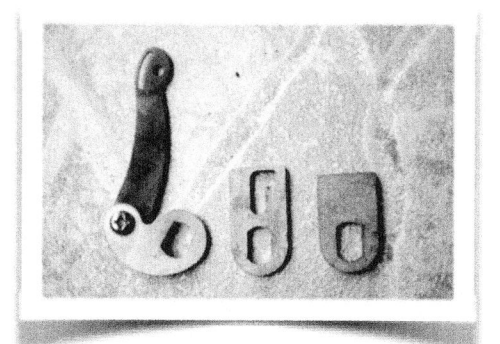

A few torque arm examples

As a rule of thumb, any time you have a system of more than 750 watts in steel or 500 watts in aluminum, you should install a torque arm. If you are using more than 1,000 watts, you should use two torque arms, one on each side. There are a number of different styles of torque arms available from a few different manufacturers, but they all work basically the same way. The idea behind adding a torque arm is to provide more material to hold the axle of the motor and resist its spinning motion. Many people make their own custom torque arms. They can be as simple as a 10mm spanner wrench (clamped or glued to the frame) to something as complex as a CAD designed and laser cut piece of stainless steel.

Torque arm installed

When tightening down the axle nuts on your hubmotor, make sure that your hub spins freely and that nothing is touching the hub to inhibit its motion. Because hubmotors are much bigger than normal bicycle hubs, some bicycles can occasionally have clearance problems. If this is an issue with your bike, try placing a washer or two on the axle inside the dropouts. This will spread the bicycle frame just a bit and may alleviate your clearance issues. If you have disk brakes, this might be another source of interference.

If there is simply not enough room between the motor shell and the disk brake caliper, you'll need to install a spacer between the disk and the motor. This can be done with 6 small washers on each of the bolts, or one larger disk brake spacer/shim.

When installing the wheel with disk brakes, it can be helpful to loosen the two bolts on the disk brake caliper that limit its side-to-side movement. This will allow the caliper to 'wiggle' freely during the installation and keep it from preventing

Installing disk brake spacer

the motor from seating down in the dropouts. Once the motor is seated in the dropouts and the axle nuts are tight, you can then retighten the bolts on the disk brake caliper.

INSTALLING THE BATTERY

Installation of your battery depends on your choice of the type of battery. Lithium batteries are generally an easy bolt on operation, but lead acids batteries will need to be installed in a bag or custom box of your own design.

The best plan for lead acid batteries is to install them as low on the bike as possible due to their significant weight. This is normally accomplished by using pannier bags or racks on either side of the rear wheel. Because of their significant weight, check and double check that you have installed them securely on the bicycle frame and that they won't budge even when hitting bumps in the road.

You may want to pack pieces of styrofoam around the batteries to help absorb the impact of bumps and pot holes. A good source for this styrofoam is the packing material that came with your motor, battery and other kit components. It often includes thin strips of foam which are perfect for packing around SLA batteries.

A wiring diagram for lead acid batteries can be seen in the battery section of Chapter 3.

With lithium batteries, installing them as low as possible on the bike will have a similar effect to increase stability, but not nearly to the extent of lead acid batteries because lithium batteries are much lighter. This means that even if you want to mount your lithium battery higher up on your bike, such as on your rack, it will only have a relatively small effect on the handling characteristics of your ebike as compared to lead acid batteries.

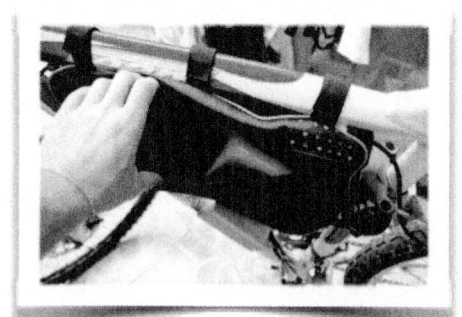

Battery in frame bag

Make sure you choose an appropriate place to mount your lithium battery based on the style of your battery. If your battery comes with an aluminum or steel plate/rail that your battery slides and locks on to, you'll want to make sure you mount this securely to the bike but without bending it, because any warping of the slide will make it difficult or impossible to place and remove the battery.

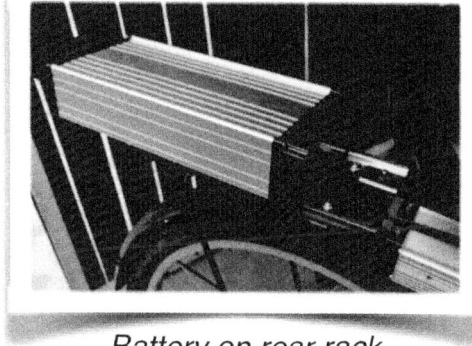

Battery on rear rack

If you don't want or need a removable battery, for example if you can bring the bike into a garage or home to charge or are concerned with theft of your removable battery, you can use cable ties or screw type hose clamps to secure the battery onto the frame or rack of the bike after mounting on the supplied plate or rail.

Be certain that wherever you mount your battery, you have enough length in the wires for them to extend to the controller. If you don't, be prepared to add wire to make it work. Refer to the soldering section if you need some help with this process.

Also, make sure that the battery doesn't interfere with the pedals or crank arms. Mounting on a rear rack means this likely won't be an issue, but for those mounting the battery on a frame member near the front of their bike, potentially inside the front triangle, this could be an issue. Spin the pedals to make sure everything clears but remember that your foot will add some length to the pedal. Be sure there is room to spare.

Lithium batteries are typically well packed inside their aluminum cases with a layer of black, high density foam to protect them from bouncing around within the case. This means that you can mount them rigidly on your bike, unlike mounting lead acid batteries which require more care.

> Quick Tip: When I use hose clamps to secure something to the frame, I like to put a piece of old inner tube between the frame and the clamp to avoid scratching the paint. I also like to put a piece of heat shrink over the free end of the hose clamp to cover any sharp burs from catching on clothing or skin.

One caveat for lithium batteries is that the vibrations from the road can cause the connectors, such as the charge connector or fuse holder, to slowly unscrew within the case over time. You'll want to periodically check these connectors. Just wiggle them between your thumb and finger and make sure they still feel tight. If they ever loosen up, you can tighten them from the inside by opening the case. Just be careful to close everything back up without pinching any wires between the case and the end cap.

My preference is to add a dab of hot glue on the back side of the connectors to keep them from unscrewing. Sometimes I do this from the start before installing the battery as a form of preventative maintenance. Make sure that you are careful when working inside the battery though, and try to use non-metallic tools whenever possible to avoid causing a short.

Mounting the controller

The controller is one of the easiest parts of the bike to mount, but before you just start drilling and bolting away, choose your mounting location wisely. You'll want to make sure that the wires are either coming out horizontally or downwards; any direction except upwards is ok, because you don't want water to run down the wires and into the hole in the controller where the wires enter. Horizontal mounting with the wires facing toward the back of the bike will help shed water, it is best to mount the wires either vertically or angled downwards whenever possible.

Some people like to mount their controller in a bag on the bicycle, while others prefer to leave it exposed for better air flow. If you have a low powered controller, a bag is typically fine because it won't be generating much heat. Also, smaller controllers usually don't come with mounting holes, so a bicycle bag is a convenient option.

Controller hidden in triangle frame bag

Most controllers larger than 350 watts will have holes on either end of the controller designed for mounting the controller to the ebike. These holes can be used to mount the controller to bicycle frame, either by drilling and screwing right into the frame or by lashing with cable ties to the bicycle's frame or rack.

You can also fabricate a plate that mounts to the backside of the controller and sandwiches a frame member

> Quick Tip: The cut ends of cable ties can be very sharp. Make sure to cut them flush to the head of the cable tie so you don't accidentally slice your leg. Trust me, I'm speaking from experience...

or rack spar between the plate and the controller. Bolt the plate and the controller together and the squeezing force will keep it firmly mounted. I recommend using nylock nuts which will not back out over time.

In the following chapter, you'll learn about all of the connections you'll need to make between the controller and the other components of your ebike. For now, just make sure that all of the wires can easily reach your controller wherever you've mounted it, and that the holes where wires enter the controller are not going to direct water into the body of the unit.

Installing the Throttle

If you thought mounting the controller was easy, wait until you get to the throttle, it's a piece of cake! Start by removing the rubber grip you have on the right side of your handlebars. Sometimes the rubber holds tightly onto the handlebar and doesn't want to budge. If you have a source of compressed air and a pointed nozzle, you can try sticking the nozzle in the hole in the end of the rubber grip and applying air or sliding the nozzle under the open end of the rubber grip. The air helps to inflate the grip slightly so you can remove it. If you don't have compressed air, you can use a sharp knife to slice down the length of the grip and cut it off.

Your new throttle likely doesn't need a rubber grip (if it is a full twist throttle) or comes with the correct matching grip (if it is a thumb or half twist throttle). Once you have the handlebar bare, slide the throttle on.

Position it so that it is comfortable to use and does not interfere with reaching for and pulling the brake lever.

Close the set screw on the side or bottom of the throttle with a 3mm allen (occasionally it is a 2.5 mm allen or even a phillips head screw). If required for your throttle, place the new rubber grip supplied with the throttle on the handlebar after the throttle is in position.

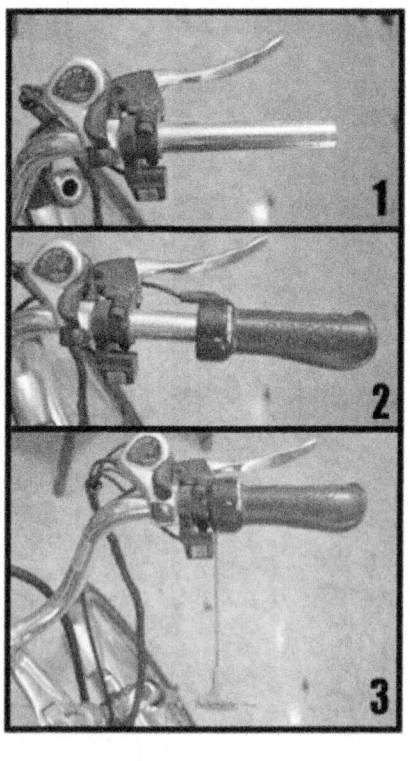

If your throttle is a half twist or thumb throttle, it likely came with some type of plastic insert or large diameter plastic or metal washer. This goes between the throttle and the rubber grip to keep them from rubbing against each other. Friction between them could cause the throttle to stick in the open position, a dangerous condition. Always install the washer if your throttle came with it. If it didn't, make sure to install the rubber grip with a small gap between the end of the grip and the edge of the throttle. You just don't want anything inhibiting the throttle from springing back to its neutral position.

The only potentially tricky part of installing the throttle can be if it interferes with the functioning of the shifter. Occasionally, shifters with levers, as compared to the twist style shifter, will interfere with the housing of the throttle or the throttle cable.

Usually this interference can be corrected by rotating the throttle up or down before tightening the set screw. This moves the housing of the throttle away from the shifter levers. Occasionally you'll need to loosen the shifter set screw (a 4mm or 5mm allen) and move it in further up the handlebar to keep it from interfering with the throttle.

If you have a twist style shifter, it may be more difficult to use with a throttle on the handlebar. This can be solved by moving the shifter to the other side of the handlebar, although this will reverse the shift pattern.

Other accessories

After the four main ebike components, the motor, battery, controller and throttle, the rest of your accessories can be installed now, or you can wait for later without affecting the primary functionality of your ebike.

Most of the accessories should be simple to install. Mirrors will usually bolt right onto the handlebar, though some are designed to fit into the bar ends. If you intend to use bar-end style mirrors, be aware that a full-twist throttle could interfere with a mirror on the right side of the handlebars. However, you'll generally want to place your mirror on the left side anyways for most drive-on-the-right countries.

Most bicycle storage devices are made to attach easily to the bike, either in the form of a basket for the handlebars and/or the rack, or a bicycle bag that simply connects with velcro.

Bells, horns, lights and other similar bicycle accessories all bolt on fairly easily and most come with quick releases so they can be removed if you are worried about someone stealing them. If anything, I believe this quick release feature just compounds the theft issue because most people don't take the time to remove their lights when they lock their bike up, meaning the quick release just makes them easier to steal. Also, the quick release is another point of potential failure. To fix this, I usually throw a dab of hot glue in the quick release mechanism. This makes it fairly permanent and keeps the lights from vibrating off if I'm riding on grass or gravel.

Upgrading the seat is also a simple process. The metal piece that clamps the seat to the seat post is called the "guts" and can be opened with either a 13mm or 14mm wrench, or occasionally an 8mm allen key. Open the nuts on either side of the bottom of the seat just enough to slide the seat off the seat post. If your new seat came with guts installed, just slide the guts from your new seat over the seat post and close. If the new seat did not come with its own guts, transfer the old guts onto your new seat and then tighten the guts to the seat post. Sometimes it can be a bit tricky to transfer the guts over and it will seem like it doesn't want to fit. Just stick with it and eventually you'll be able to coerce it into place. Sometimes I put one side on the first rail under the seat, then rock the other side down onto the rail. Leave the bolt out until you've got both sides on the rail, then slide the bolt through the center. The first time is always tougher, but it will be easier if you ever have to do it a second time!

You'll more than likely want to adjust the angle of the seat to match your comfort preferences. Try tightening the seat down first and then sitting on the seat to check the angle. If it isn't right, loosen the nuts a bit on either side of the guts and adjust the angle before retightening and rechecking.

Chapter 6: Making the Connections

Now that you've got all of your components mounted on the bike, it's time to do something with all of those wires hanging on the floor. Start by loosely gathering the wires from the handlebars, such as the wires from the throttle, ebrakes, headlight, and any other accessories you added to the handlebars.

Cable ties will be your best friend when organizing the wires. Use cable ties to group the wires and cables together in a neat fashion along one side of the bike. You may want to follow the brake or shifter cables to keep all of your wires neat looking and in one place. I like to use spiral wound plastic to make all my wire bundles neat and professional looking. You can also use electrical tape to get a similar look, though it is more annoying to remove later if you need to make changes.

As you bring the wires back towards the controller, follow along the frame and install cable ties at regular intervals in order to make sure the wires are securely mounted.

Make sure that you have left enough slack in the bundles by turning the handlebars to the extreme positions to the right and left. If the cables are too short and limit the movement of the handlebars, or become kinked or otherwise stressed as the handlebar turns, allow more slack in the cables or reroute them accordingly.

The motor wires should be handled with extra care, because any stress to these wires can cause big problems and be difficult to fix. There are two schools of thought on how to best secure the wires exiting from the motor. The first is with the wires exiting down, creating a "drip loop" that allows water spray to drip harmlessly down onto the ground and not follow the cables back into the motor. The second is with the wires exiting up and secured tightly to the frame immediately after exiting the axle. This ensures that if the bike ever falls on its side, the wires are not cut by the axle because they are below flush with the tip of the axle.

I like to combine these two approaches by making a tight drip loop that still pulls the wire back against the frame and below the end of the axle (see the picture below).

The drip loop method works better in rainy areas while the straight up method is better suited to dry areas. Either way, make sure to secure the wires against the frame so they do not get caught on anything while riding. I mentioned this previously, but considering the money invested in your motor, it bears repeating: breaking a motor wire near the axle can require a complicated and difficult repair. Don't let this happen to you!

Continue securing the motor wires up the frame with cable ties until they reach to the controller. If you purchased your parts as a kit, the connectors should perfectly match each other. Simply plug the connectors from the motor, throttle and other accessories into the matching connector on the controller.

Some controllers have numerous connectors, so searching for the connector with the matching number of wires can help you find the correct one. A three-wire throttle connector will only match with another three-wire connector on the controller. Secure any extra wires in a bundle with cable ties. I like to then take a piece of old bicycle inner tube or a scrap of black leather and wrap it around the bundle of wires coming out of the controller. This keeps everything looking neat and organized instead of resembling a rainbow-colored rat's nest of wires. Black electrical tape also works for this, but is more annoying if you need to remove it later to change a component or search for a problem.

If your battery came with the appropriate connectors for your controller, simply plug it in as well. However, many times the battery will require you to make a custom final connection between the supplied battery connector and the controller. See the soldering section below.

If your parts were purchased from multiple sources and/or the connectors don't match the controller, you'll need to make your own connections either by swapping out some connectors or cutting off the connectors altogether and soldering the wires directly together. This is not complicated, but will take some time to make sure you have secure connections that won't break, collect water and short out, or reduce the path of the electrical current.

Soldering

There are a number of great sources available to teach you how to solder wires. Therefore, I'm going to give a simple overview and provide some resources at the end of the chapter to provide additional assistance.

To begin, you'll need the following tools: soldering iron, soldering iron holder, solder, and a wet sponge or brass sponge. You may also want to use a device called a "helping hands" to hold your work for you if you are working on a surface. Helping hands are two small, moveable wire clips that hold the wires steady as you solder. They work better for bench work than if you are soldering wires already on your bike. Heat shrink tubing is helpful as well, but if you don't have access to heat shrink tubing you can use electrical tape instead.

I much prefer a variable temperature soldering iron station because I can adjust the temperature depending on the size of the wires and type of solder I'm using, but a simple standalone soldering iron will work. I just get frustrated by those low power soldering irons because I have no control over the temperature and therefore less control of the final quality of the soldered connection. To help you out, I've listed a couple of options for soldering stations in the resources section of this chapter.

Soldering is the process of using molten metal to securely bind together metallic contacts for electrical connections. For our purposes, most soldering will be used to join two wires together, so this is the process that I'll describe here. To begin, prepare your wires by stripping the plastic insulation from about 1/8" to 1/4" from the end of the wires.

Next, secure the wires so they don't move while you're working on them. You can use a helping hands, or simply place something heavy on the wire near the end to hold it. Make sure to elevate the end of the wire so you can manipulate it without the work surface getting in your way.

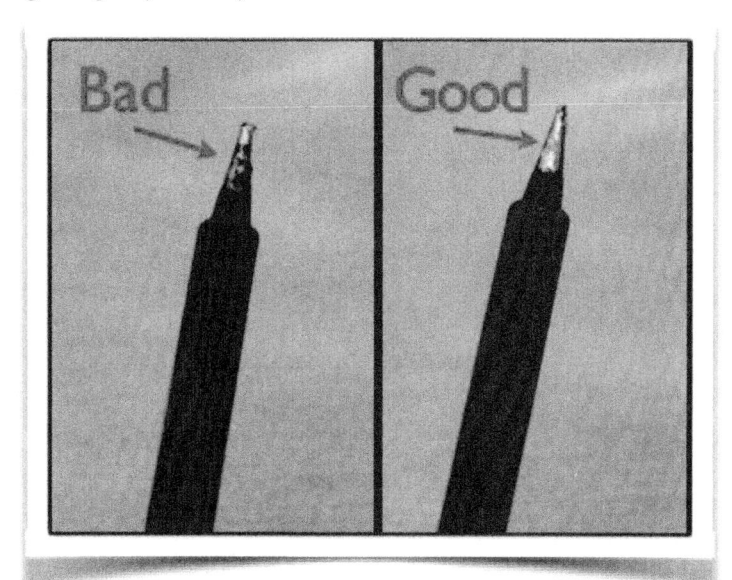

Left: oxidized soldering iron tip, not good for heat transfer; Right: cleaned and tinned soldering iron tip

Turn on the soldering iron and allow it to heat up. Check the recommended temperature for your solder; it is usually around 600-700 degrees Fahrenheit (315-370 degrees Celsius). Once your soldering iron tip is hot, 'tin the tip' by placing a small amount of solder on the tip and allowing it to melt.

Now, clean the tip using a wet sponge or brass sponge. A standard sponge needs to be wet to keep the soldering iron from burning it, but a brass sponge is used dry. I prefer a brass sponge, because I never have to worry about wetting it. It is important to keep the tip of your soldering iron clean to create a good surface for heat transfer.

Now, tin the tip a second time with a small amount of solder. This helps create even heat distribution on the tip of the soldering iron.

Hold the tip of the soldering iron against the bare wire that you'd like to solder. The hottest part of the iron isn't actually the very tip, but a few millimeters back from the tip. Allow the iron to heat the wire for a few seconds, and then apply some solder to the point where the iron is touching the wire. You want enough solder to cover the wire, but not so much that it starts dripping off.

Let the soldering iron tip remain in contact with the wire for another couple of seconds while it heats the solder on the wire tip. Remove the soldering iron and let the wire cool for a few seconds until the solder hardens. The wire is now tinned with solder on the bare end. Repeat this process for the second wire.

If you have heat shrink, cut a piece approximately 1 inch (2.5 cm) long and slide it over the end of one wire. Now bring the two wires together so that their soldered ends are next to each other. Align them so they lay side by side, not tip to tip. This will give more room for the solder to securely bond them together.

Hold the soldering iron tip against the two wires at the connection until you see the solder melt. If you have enough solder on the wires originally, you won't need to add more. If the solder doesn't seem to melt well or doesn't join the two wires together, add more solder and try again.

Once the wires are joined together by the melted solder, remove the soldering iron while taking care not to bump the wires apart. The connection should cool in a few more seconds. Test the connection by gently pulling on the wires. If they separate or the soldering joint cracks, the connection is not good enough and you'll need to redo it.

If the connection holds and appears strong, slide the heat shrink over the connection and apply heat with a heat gun. The heat shrink will shrink and form a barrier over the solder connection to prevent short circuits as well as resist water incursion. If you don't have a heat gun, try a hair dryer on high heat. Also, a lighter will work in a pinch, although this is not the preferred method and you should be careful not to damage the protective insulating sheath on the wire near the heat shrink.

Sometimes it can be tough to hold two wires together and hold the soldering iron. When I'm soldering wires on an ebike where I can't easily use a "helping hands" because the wires are on the bike, I often allow one wire to hang freely, use my left hand to hold the second wire up to the first wire and hold the soldering iron in my right hand. This allows me to deal with three objects while only using two hands (and the benefit of gravity).

If you followed my tip on placing solder on both wire ends first, then you won't need to apply solder again when bringing the two wires together, which reduces the amount of things you need to hold in your hands at one time.

Using connectors

Soldering two wires together ensures a solid connection, but you may want to use a connector in situations where you want to be able to separate the connection, such as with the battery. In this case, you'll need to follow the specific instructions for the type of connector you are using.

The good news is that most connectors follow the same basic layout. You crimp the bare end of the wire being connected into the connector's metal contact and then slide that contact into the housing. There is usually a pin on the contact that will 'snap' into the housing to secure the pin into the housing.

Many connectors require the use of a special crimping tool, designed to work with their specific contacts. If you don't have access to the special tool, you can usually get away with crimping the connector with a pair of needle nose pliers and then using a drop of solder to secure the connection. You'll want to use the solder because the crimp won't be as strong without using the specific tool.

Some connectors require you to place the housing on the wire first, then place the contact on the wire. The housing then slides on from the backside to cover the contact. You'll need to verify the instructions for your specific connector.

There are some connectors that have become common on ebikes due to their ease of use and good performance. Anderson PowerPoles are rated up to 45 amps and are good for connecting phase wires from the motor or battery leads. For chargers, many people use RCA or XLR connectors. These are good for handling up to 5 amps, the maximum power level commonly used by chargers.

Connecting everything to the controller

The controller has many wires and connectors that allow you to connect the main ebike components, as well as a number of accessories. There are 4 main sets of connectors that will be crucial to your ebike's functionality: the battery, throttle, and two sets of motor connectors (three thick wires and five thin wires). The rest of the connectors allow additional components that may be of interest to you, but are not essential for the operation of your ebike.

Battery connection

The first important connector on your controller is the power, and this will be in the form of a large red wire and a large black wire. As you would imagine, red is the positive and black is the negative.

Occasionally you'll see a small wire, usually red, but sometimes orange or purple, that shares the same connector with the large red and black power wires. This smaller wire is the enable/disable wire and basically works like an ignition switch. When it is connected to the large red wire, the controller is effectively "on", and when it is disconnected from the large red wire, the controller is effectively "off".

If you have a switch somewhere on the bike, usually on the throttle or headlight, you can wire the switch wires with one wire going to the large red power wire and the second switch wire going to the small enable/disable wire. When you turn the switch, the enable/disable wire and the power wire are connected or disconnected, turning the controller on or off.

If you don't have a switch on your ebike, just wire the enable/disable wire and the positive red power wire together permanently. This way, when you connect the power from the battery (either by turning the battery's own switch or by connecting the battery connector), the controller will already have the enable/disable wire in the "on" position. This means that when the battery is connected, the ebike will be instantly "on".

Throttle connection

The next step is to locate the throttle connector. This is usually a 3 wire connector. One wire will be positive (always red), one wire will be negative (nearly always black but sometimes blue) and one wire will be the signal (usually green or white, rarely blue).

If you bought your ebike parts as a kit, your throttle will probably have a matching connector for the controller so the connection is simple. If not, you'll need to wire it together yourself. You can use a three pin connector, but a better connection can be made by soldering the wires directly to the controller wires. The connection will be stronger, but you won't be able to quickly disconnect it to change throttles like you can with a connector. For most people this won't be an issue as you rarely, if ever, need to change your throttle. In the event that you do want to change a throttle, you can simply snip the wires at the solder joint and replace the throttle with a new one.

If your throttle has more than three wires, you'll need to determine the role of each wire. If you have a button or switch on the throttle, two of the wires will be connected to it. To determine which two wires connect to the switch, use a multimeter on the continuity ("beep") setting. Touch the probes together to make sure that the multimeter beeps when it senses continuity in a connection.

Now, while the throttle is disconnected, test between each wire of the throttle to determine which two wires are connected by the button or switch. You can eliminate the three main throttle wires (usually red, black and green/white). For many of the most common throttles, the switch is connected to brown and yellow wires, but the color of the switch wires can vary from throttle to throttle. Green is sometimes used as one of the switch wires too.

If you don't see a connection between any wires, try turning the switch to the opposite position, or pressing the button one more time, as it may have been in the 'off' position. Once you've located the two switch wires using the beep setting on your multimeter, you'll want to wire them to whatever you plan to control with the switch, usually the enable/disable wire from the power leads on the controller. The enable/disable wire is the thin red wire running along side the thick red power wire, which we talked about in the controller section.

If you don't have an enable/disable wire on your controller, but still want to be able to use the switch on your throttle, you can wire the switch wires inline with the black negative wire for the throttle. This means that when you turn off the switch you will break the circuit sending power to the throttle, keeping it from working. Technically, your ebike will still be on and drawing a small amount of current from the battery, but your throttle will be disabled. This would be a good way to disable your bike if you were just parking outside of a store for a few minutes or stopping at a street kiosk and didn't want to go into your battery bag to disconnect the battery connection. This would not be a good way to disable your bike if you wanted to leave your bike for weeks or more, because it can slowly drain your battery.

If you have LEDs on your throttle, there will either be one or two wires to feed power to the LEDs. If there are two wires, one is positive and one is negative. If there is only one wire, it is positive (and the negative of the LEDs is shared with the negative of the throttle, almost always the black wire, though very occasionally blue).

Wire the positive wire to the battery positive. If your controller has the enable/disable wire, it is better to wire the positive LED wire together with the small enable/disable wire. That way, when you close the switch, the LEDs also turn off.

If you do the opposite and wire the LED positive with the large battery positive wire, the LEDs will be lit anytime the battery is connected, even when you've closed the switch to the enable/disable wire.

Motor connection

Your motor should have two sets of wires coming out of it, three large wires and five small wires. The large wires are called 'phase' wires and carry the electricity that powers the motor. The smaller wires are called 'hall' wires and connect to the hall sensors inside the motor that keep track of how the motor is spinning.

If your motor is sensorless, it will only have the three thick wires, making connections easier. These motors are more rare, as using sensors is the standard.

If your controller and motor came together as part of a kit, you will likely have matching connectors that you simply need to click together. It is possible that the motor came with pins installed on the hall wires but with the connector housing separate. This is to help you change hardware on the axle, if needed, because the nuts won't fit over the plastic connector.

When you are ready to connect the motor to the controller, you will simply have to push the pins into the connector housing. Look at the connector on the controller to match the colors of the wires in the correct positions.

If your motor did not come with your controller and you purchased them separately, the process can be a little more involved. Not only do you need to make your own connections manually, but you may need to play with the order of the phase and hall wires to find the right combination. This is because with three phase wires and three hall wires (the small red and black wires are power wires to the hall sensors), there are actually 36 different possibilities for connecting the wires.

By changing the order of the wires, you change the order that the phases in the motor function and the order in which the hall sensors interpret the motor's spinning. If your motor and controller came from different sources, you'll need to experiment with the phase and hall order until you find the right one. This can be a fairly slow and annoying process, but the good news is you'll only have to do it once. Once you've found the right order, it will be set for the life of the motor and/or controller.

There are three rules to guide you as you are switching motor wires to find the right combination. The first rule is that thick wires only go with thick wires and thin wires only go with thin wires. The three blue, green and yellow thick wires can be switched among themselves, but you should never connect the thick wires and thin wires together, because they carry different currents and go to different places within the motor. The second rule is that any time you are switching the order of the wires, you should have the battery turned off. It can be annoying to keep switching the battery on and off, but it will help prevent the possibility of causing a short circuit, which could damage your components or in rare cases cause injury. The third rule is that the red and black thin wires never get recombined; red always goes to red and black always goes to black. Always.

Some people just start randomly switching wires until they find the combination that works. This is a poor method that will take a very long time and frustrate you to no end as you try to determine which options you've tried and which are left. More than likely you'll end up repeating combinations that you've already tried.

I've had to do go through this trial and error process of finding the right hall and phase order more times than I can count. To keep you from having the same negative experience, I've developed a pattern that

I follow to make the process as efficient as possible. To find the right connection, I use the following trial and error sequence:

Trial 1: colors matching;

Trial 2: only blue matching (switch yellow & green);

Trial 3: only green matching (switch yellow & blue);

Trial 4: only yellow matching (switch green & blue);

Trial 5: nothing matching #1 (all mismatched);

Trial 6: nothing matching # 2 (all mismatched but shifted one spot).

This is the pattern I use to make sure that I don't accidentally repeat any combinations and to remember what I've tried and what is left to try. It might not immediately make sense, so let me explain it a bit more. First, remember that the thin red and black hall wires will always connect with the controller as red to red and black to black. You should go ahead and solder/shrink wrap these to get them out of the way. Also, they'll need to be connected for the remainder of the test sequence to work.

Next, use my sequencing to test the connections in this manner: wire the three large phase wires together as well as the three small hall wires, colors matching. This is my "Trial 1". I like to use alligator clip test wires to do this because this method requires no soldering and makes changing the combinations quick and easy.

Make sure you don't allow the test leads of the alligator clips to touch and make a connection. I like to slide small pieces of cardboard between them to be safe.

Now that you've got everything connected as in Trial 1, turn on or connect your battery and give a small amount of throttle. Make sure that the bike is either upside down or somehow has the motor wheel elevated off the ground so it doesn't drive off. You can also lean the bike on its kickstand to elevate the motor wheel a few inches to perform the test.

If this combination of wires is correct, the wheel will spin nicely, in the correct direction, and make very little noise. If the motor is a direct drive motor, most of the noise coming from the properly functioning motor should be the wind noise of the tire. A properly functioning geared motor will be a little louder due to the gear noise but will still sound smooth. If you hear grinding, crunching, high pitched squeals or otherwise bad sounds, and/or the motor shakes, jumps irregularly, skips or doesn't move, you know the connection order is wrong.

Now, assuming this first try was not the correct order, turn off the power and try Trial 2 on just the phase wires (the thick wires). Remove the yellow and green phase wires and switch them, leaving the blue alone. Don't change the small hall wires. Turn on the power, give a little throttle again and check. Again, bad noises, jerking or no movement at all indicates that the order is still wrong.

Next turn off the power and try Trial 3 with the phase wires with only the green phase wire connected color to color and switching the blue and yellow. Don't touch the thin hall wires yet. They should all still be color to color. Turn on the power and check. If that order isn't successful, keep trying Trials 4, 5 and then 6. If none of them are successful, now is the time to change the hall wires.

The hall wires should have been in the Trial 1 configuration the whole time, with each of the three wires (blue, green and yellow) connected color to color (and the red and black should be connected color to color as well, but you'll leave those alone). Now, change the thin hall wires to Trial 2 by switching the green and yellow hall wires. Also, change the phase wires back to the first order so they are all matching color to color. This means the phase wires are in the Trial 1 configuration and the hall wires are in Trial 2 configuration.

Now, repeat the test from Trial 1 through Trial 6 on the thick phase wires, while leaving the halls in the Trial 2 configuration. If you still haven't found the correct combination after the six tries, change the hall wires to the Trial 3 configuration, and repeat the test with the phase wires changing from Trial 1 to Trial 6.

Continue testing in this way, changing the phase wires through the six trials in order, then switching the hall wires to the next trial configuration and repeating the six phase wire trial tests. If you continue through all the options, you will have tried a total of 36 different wire combinations.

Based on the way the motor is designed, there are actually 3 different configurations that will make the motor spin correctly in the right direction, and 3 different configurations that will make the motor spin correctly but in the wrong direction (i.e. driving backwards). You'll just need to keep trying options, sticking to the sequence that I have described, until you find one of the three correct, forward spinning options.

If you have gone through all 36 options and have been unsuccessful, you have a bad connection somewhere. Check the positive and negative hall wires to make sure they are connected well. A break in the connection there can mean all 36 tests will fail.

Once you've found the correct option, connect the wires to the controller by either soldering them and covering them with heat shrink (better) or electrical tape (not as good, but still works). Test to make sure the motor still spins smoothly when you give it some throttle. If you suddenly hear bad motor grinding noises, you probably have a bad connection. Check your solder joints and redo them if necessary.

Charger connector

If you are using a lithium battery, the battery should have come with a charger with a matching connector. If you built a pack using lead acid batteries, you'll need to find a charger of the correct voltage (24V, 36V, 48V, etc.) as well as the matching connector.

If for some reason your lithium battery did not come with a charger, or the charger's connector does not match your battery's charge connector, you'll need to change the connector on the charger.

Here's how to do it. First, make sure the charger is NOT plugged into the wall or the battery. Next, cut the old connector off using wire snips. I recommend leaving at least 3 or 4 inches (10 cm) of wire left on the connector in case you ever want to wire it to something else.

Next, separate the leads of the charger wire so they are not touching each other and strip the tip of the insulation off the wires so you can see the bare copper.

Now, VERY carefully plug in the charger while making sure the leads do not touch each other, as they will be live (conducting electricity) and will create quite a spark if they touch.

Use a voltmeter to determine which lead is positive and which lead is negative. If the voltage shown on the voltmeter is negative, you have the leads backwards. If the voltage is positive, the leads are correctly wired with the red lead to the positive wire of the charger and the black lead to the negative wire of the charger.

Now unplug the charger from the wall, being VERY careful not to touch the charger's bare wires together. Remember (or better yet, write down) which wire is positive and which is negative. Most chargers I've worked with use blue and brown wires for the positive and negative, respectively, but it can vary among different chargers.

Now, using a voltmeter, test the battery's charger connector to determine which side is positive and which is negative. This will tell you which side of the charger connector to solder to which charger wire. It may help to plug in the charger connector to the battery connector even without any wires on the charger connector. This way you can confirm which pin is positive and which is negative without trying to "flip the image" of the two connector faces in your head.

When you have determined the required polarity of the charger connector, solder the wires from the charger to the appropriate pin on the connector. It is a good idea to use heat shrink to keep the wires from touching each other on the connector and shorting out, but even electrical tape will do as a substitute.

Before plugging in the charger, verify one more time with the voltmeter that you have the polarity correct with positive and negative on the appropriate pins of the connector. Once you have confirmed that the polarity is correct, plug the charger in to your battery to make sure it works. If your charger lights accordingly (with a red LED indicating that the battery is charging), you have successfully changed your charger's connector. If the charger doesn't work, you may have flipped the polarity. Check and repeat.

Chapter 6 resources

Soldering tips and tricks: https://www.sparkfun.com/tutorials/213

My favorite soldering station: http://www.amazon.com/Hakko-FX888-FX-888-Soldering-Station/dp/B004M3U0VU

A more budget friendly soldering station: https://www.sparkfun.com/products/10707

Anderson PowerPole connectors: http://www.amazon.com/Power-Pole-Connector-Black-Anderson-Sermos/dp/B000QUZD4W

Chapter 7: Finishing Touches and Maintenance

Congratulations! If you've carefully followed my instructions up to this point, you will have now successfully completed your first ebike. Take a step back and admire your work. Pretty cool, huh? Now I'm sure you're very excited to hop on and go tear up some rubber on your new ebike, but there are a few more items that you need to check and make sure are in good working order before you get to experience the fruits of your labor.

Safety check

First and foremost is your safety. You need to check all the systems on your ebike to make sure they are all still working properly. Start with the pedals and freewheel on the motor. Spin the pedals and crank arms to make sure your chain is in line, doesn't get caught between the frame and the rear gears, and that you have properly reinstalled the crank arms if you removed them to install a pedal assist system. Your pedals and chain should work just like they did before the ebike conversion.

If you find that the pedaling is jerky or irregular, check to make sure the chain isn't getting caught against the side of the rear hubmotor or pinched by the gears during the installation.

Next, check your brakes. If you are using rim brakes on a used bicycle, consider putting on new brake pads. You'll find that your brakes get much more wear when you're stopping from higher speeds. Plan on going through brake pads faster than on a regular bike.

Test the front brakes by holding the front brake down while pushing the bicycle forward. You shouldn't be able to move the bike, and pushing hard enough should either make the tire slide on the ground or the rear of the bike lift up. Now test the rear brake by holding the rear brake lever and pulling the bike backwards. With enough pulling force, the bike should either slide on the tire or the front of the bike should lift up.

When you pull the brake levers, the levers should not move back far enough to touch the handlebars. If they do, you'll need to loosen the cable clamping bolt on the brake arms near the wheel and pull more cable through. This will shorten the amount of cable you have available to pull and reduce the stroke length of the brake levers.

For disk brakes, do the same brake check by pushing the bike while holding the brakes. If you need to adjust the brakes, loosen the bolt that clamps the cable on the bottom or backside of the disc brake. Pull the cable slightly out and retighten. Do this as many times as needed until you achieve a cable length that provides strong braking while still letting the wheel spin freely with no brake interference when you release the brake. You can also make a fine tune adjustment on the brake pad closer to the centerline of the bike by turning the allen bolt on the inside caliper. You'll need to access it by sticking your allen wrench through the spokes from the opposite side of the wheel.

Check all of the nuts and bolts that support weight-bearing parts of your ebike like the wheels, rack, headset, handlebars, seat post, etc. It is important that these are all torqued tightly. When in doubt, use a little blue Loctite or similar thread sealer to keep these nuts and bolts from backing out due to vibration.

Safety is the number one priority in building any vehicle, especially when you've done it yourself for the first time. Everyone has to learn how to do something by trying it on their own the first time, but that doesn't mean you can't get a little help. There's no shame in bringing your ebike to a local bicycle shop to have them check out your brakes, cables and other working components just to make sure your bike is as safe as possible and road ready. You might even impress them with your skills and begin to set up a relationship if you ever need their help.

I'm also listing here a number of links to great resources on general biking safety, all of which apply, and are even more important now that your bike has been upgraded into an ebike capable of much higher speeds.

http://bicyclesafe.com - some great articles (including a straightforward headline "how to not get hit by cars").

http://www.be-safe.org/css_com/bicycle/ - interactive bicycle safety website that allows you to follow through at your own pace, when you have time. Just click continue to begin.

http://www.nhtsa.gov/Bicycles - This one is a great example of your hard earned tax dollars at work.

http://www.helmets.org - a LOT of info on helmets and bicycling. This one is a no-brainer (get it?!)

TEST RIDE

When you're ready for the maiden voyage of your ebike, remember to have fun but to be safe. Take it easy and remember that you're going for a test ride, not a qualifying lap at Indy. Start out by listening for any noises coming from your bike that could indicate a problem. A common 'tick tick tick' noise is usually something rubbing on the wheel somewhere or a loose spoke dancing around. Check for interferences of the tire on the frame, a wire on the tire, brake pads on the rim or disc, etc. You'll need to track down the source of any noise and make sure you correct it. It is usually as simple as installing a few more cable ties to secure a dangling wire or making a small brake adjustment. Of course, you'll need a few tools with you but nothing major.

You'll surely want to determine your range to understand how far you can go on one charge, but its better for the overall health and longevity of your battery if your first 3-5 rides only partially discharge your new battery. Try to only ride 5-10 miles for your first few trips and then charge the battery. After you've done a few smaller charge cycles, then you can test just how far your battery will go before it reaches the low voltage cutoff and you have to pedal home. This is a good piece of information to know, but discovering it isn't the healthiest thing for your battery. Lithium batteries can definitely be run all the way down to empty, but you should try to minimize how often you do it.

As a rule of thumb though, try not to completely drain the battery too often, especially if you are using SLA batteries. Completely draining a battery can reduce its overall life expectancy. Every now and again isn't a problem, but running your battery dead everyday could cut its useful life by as much as half (or more for SLAs).

Maintenance

For the first few times you ride your ebike, check all the components upon returning home to make sure all are in good condition and working within specs. Feel how warm your motor and controller get. Check all of your wires to make sure they are placed well and aren't getting rubbed by the pedals or tires. Just make checking your ebike a regular habit over the first few weeks and months you are enjoying it.

In fact, making a habit of these routine inspections will serve you well in the future as you may see a potential problem arising before it has the chance to leave you stranded on the side of the road or worse, result in injury.

One of the most common problems that occur in ebikes due to lack of maintenance is a wire connector failure, where the wire either backs out of the connector or breaks off completely near the connector. Routinely checking your connectors can help avoid this issue. If you see what looks like a weak connection, fraying wire or broken insulation on a wire near a connector, consider replacing it immediately. Securing your wires well from the beginning can help avoid this problem down the road. Vibrating and dangling wires are just asking to break due to fatigue.

Another important area to watch are the spokes on your wheels, especially if you are using a hubmotor. They can loosen up over time and this can result in broken spokes and/or forcing the wheel out of true. If you try to squeeze two spokes together, they should be fairly firm and resist bending (but not too much - you should still be able to bend them a bit by hand when you squeeze two together). If a spoke is very easy to bend, it's too loose. If it rattles, it's definitely too lose.

Tighten any loose spokes with a spoke wrench until it matches the other spokes for stiffness, or bring your bike to a bike shop and ask them to true the wheel. Over-tightening a spoke can result in pulling your wheel out of true, meaning it won't be perfectly round when it spins and is likely to interfere with your brakes. If you don't feel comfortable adjusting your own spokes, any bike shop can do it for you. You should check your spokes after the first 50 miles, 100 miles and 200 miles and then every 200 miles after that to confirm they are all still tight and haven't loosened up. Generally they should be fine, but if you do notice any loosening, take care of it quickly so it doesn't become worse.

As I mentioned earlier, brake pads will require frequent checking and replacing. A daily commuter ebike could need new brake pads every couple months, while an ebike that rides less frequently might get away with six months or more before needing new brake pads.

You can check the state of wear of your brake pads by looking at the wear line on the top of the pads. This will be a visible mark that runs around the circumference of the pad. If the pad has worn down close to the line, it's time to change them. If you hear scraping noises coming from your brakes when you stop, it's definitely time to change them. That's the metal from behind your brake pads that has worn through and is now scraping directly on your wheel rims. Left unchecked, this can destroy your rims in a hurry.

Disk brakes, especially hydraulic disc brakes, are harder to check and change by yourself and should be taken to a bicycle shop for maintenance if you notice decreased braking performance and don't feel comfortable adjusting or checking them yourself.

Chapter 8: Sample Ebike Builds

If you've read Chapters 1-7, then you've already received all the information you need to be able build your first ebike. In those pages you've taken in the basic knowledge, the sources for parts, and the instructions to assemble your complete ebike. But there's one thing more that I haven't been able to provide to you with: experience.

One of my favorite quotes is by Mark Twain, who said "knowledge without experience is just information". So far, you have acquired information. To be fair, it's a lot of very detailed information, but it's still only information. And there's no substitute for experience.

While the idea of building your own ebike may still seem daunting to you, for me, an experienced builder and designer, it's just an hour's worth of enjoyable work. What's the difference? I've built hundreds of ebikes and that experience is what makes it so easy.

It's hard for me to directly impart to you that kind of confidence inspiring experience, but I hope that I can at least help take you along on a few sample, 'on paper' builds that will help you further understand how easy it is to use the knowledge you've gained from this book and turn it into an actual ebike. So let's dive right in and look at three very different, theoretical, ebike conversions.

Charlie's commuter

Charlie currently drives or takes the bus to work in the city everyday, a 12 mile (19 km) round trip commute. He wants to build an ebike that he can ride everyday to work and back on one charge. He'll charge it every night at home, but because he lives in an apartment, he'll need to take the battery inside to charge while leaving the bike locked outside.

He wants the motor to be powerful enough to commute with, but Charlie doesn't have too many hills to deal with and only wants a top speed of 20 mph (32 km/hr), enough to keep up with city traffic. He wants a bicycle with standard size 26" wheels and doesn't need suspension because he plans on staying on flat roads and sidewalks, not going off-road. He is pretty good with his hands and working with tools and so he doesn't mind assembling his kit from a few different vendors, even though it means he might need to do some custom modifications to the connectors.

Charlie starts by finding a suitable bike for his build. Because he doesn't need suspension, this opens up numerous excellent options for him in a reasonable price range. He decides to go with a $100 used Giant bicycle he gets from Craigslist. It has a steel frame that he chooses for the strength of steel over a newer bike with an aluminum frame.

Buying a used bike also saves him some money that he can now spend on a nice lithium battery. He knows he wants a 36V system and a direct drive motor, so he goes with a Nine Continent 2807 rear motor kit from www.Ebikes.ca (item code NC26RD), which includes the motor already laced into a 26" rim, a 20A controller, a throttle and a Cycle Analyst for viewing all the motor and battery information during a ride.

He inputs the battery, motor and controller values into the Ebikes.ca hubmotor simulator and sees that the combination of parts that he has selected should give him a top speed of about 21 mph (34 km/hr) which is just about spot-on his original target.

For a battery, Charlie chooses a 36V 10AH Ping Battery from www.PingBattery.com and picks up a large triangle frame bag off Amazon.com to mount the battery inside the front triangle of his frame.

He receives all the parts in the mail and begins installing them on his bike. He opts for new Maxxis Hookworm tires combined with Joe's tube sealant to limit the likelihood of flat tires. He removes his old rear wheel and replaces it with the new Nine Continent motor wheel with the Maxxis tire.

Because his bike had rim brakes, he doesn't have to worry about adjusting disk brakes with the motor.

His battery goes in the frame bag and he uses Anderson Powerpole connectors to have a strong, quick release connector between his battery and the controller.

His controller, which came with the kit from www.ebikes.ca, is mounted on the frame tube where a water bottle cage often mounts. He used one bolt of the water bottle cage to bolt through the mounting hole on the controller and used cable ties on the other end of the controller to secure it.

His throttle mounted quickly and easily on the handlebars, and the Cycle Analyst mounted just as easily. All the wires are cable tied to the frame.

Charlie's bike proves to drive at about 20 mph (32km/hr) on flat ground, slowing down to about 15 mph (24 km/hr) on slight to medium sized hills. He finds that he has about a 20 mile (32 km) range without any pedaling, and up to about 35 miles (55 km) with pedaling. This is more than enough for his daily commute to work and back.

Every night he locks his bike up with a Kryptonite "Fahgettaboudit" lock on the frame as well as a smaller cable lock through the hubmotor and undoes the velcro on his battery bag to bring it up to his apartment to charge.

Total cost including approximate S&H:

- Bicycle (used): $100
- Motor kit: $550
- Battery: $510

- Tires and sealant: $100
- Locks: $115

Total: $1,375

Sally's slowpoke

Sally wants an electric bike that she can use for driving around town for fun and occasionally running small errands. She plans to do mostly sidewalk and bicycle path riding and so she doesn't want something too fast or powerful, just a relaxing and easy to use ebike.

She is also on a budget and wants to try to keep the cost down as much as possible. She has her eye on an ebike that can travel at about 15 mph (25 km/hr) and is easy to pedal manually if she wants the option.

To save money, she'll also be using a Dahon folding bike with 20" wheels that she already has in the garage. She gets it checked out by a bicycle shop first to make sure it's in good shape, including having the brake pads and cables replaced and a thorough cleaning and oiling of all the moving parts.

Sally decided that she wants to go with a Chinese supplier over a North American one to save money. After shopping around on the internet, Sally chooses to go with a Q100 36V geared front hubmotor kit from www.BMSbattery.com. She enters the option for a 20" wheel and chooses the 328 RPM motor, which should give her just over a 19 mph (31 km/hr) top speed.

She chooses a 36V 10AH Lithium "little frog" battery from the same site. The parts arrive and she begins the installation process. She swaps the tire, inner tube and rim strip from her old front wheel onto the new motor wheel and bolts it into place in the front fork.

Her battery included a mounting rail that bolted right onto the seat post. The unit had room for her controller as well, so she placed the controller inside the mounting rail box and used electrical tape and cable ties to hold it in place and keep it from rattling around. Conveniently there were already holes in the bottom of the battery mounting case's plastic cover for wires from the controller to exit.

The kit came with a connector for the battery wires to connect to the controller, but she had to put the connector on herself. She crimped and soldered the positive and negative leads from the battery rail to the included pins, and then pushed them into the connector until she heard a click, making sure they aligned with the correct colored wires on the controller's connector.

The throttle and LED battery display both mounted on the handlebars and she used cable ties to secure the wires down the handlebar stem along with the motor wires. Sally then snaked all three wires back to the controller on the seat post.

Because she bought a kit with all the parts together, the connectors all matched up, including the wires for the motor, so connecting everything was a snap. She picked up a few accessories as well from her local bicycle shop, including a bicycle light, lock and helmet.

After finishing all her connections and securing all of the wires, Sally finds that her new ebike has a top speed of about 18.5 mph (30 km/hr) which is a little faster than her 15 mph (25 km/hr) target, but she is enjoying the little bit of extra speed when out on the wide open bike trails. Her range is about 16 or 17 miles (25-27 km) without pedaling, but she finds that pedaling greatly increases her range. She also enjoys the fact that pedaling with the motor turned off is easy, because she chose a geared hubmotor, which has negligible forward rolling resistance.

Sally also finds that the frog style battery is very easy to remove and take inside for charging, which is a benefit considering her range is in the middle to low end of an average ebike's range. All in all, her project turned out to be a exactly what she was looking for, A small, comfortable, affordable yet fun ebike.

Total cost including S&H:

- Bike: free (she had it already)

- Motor kit with controller and battery: $515

- Tune-up and bicycle accessories, lights, helmet, lock etc: $100

Total: $615

Ron's racer

Ron has a need for speed and wants a fast ebike. He isn't going to use it for groceries and he's not going to use it to go to work. He's going to use it to ride fast and have fun. His goal is to be able to do at least 30 mph (48 km/hr) and have a comfortable ride. Money isn't a huge issue. He also wants to be able to ride both on and off-road, so a full suspension bike is a must.

Ron goes to a local bicycle shop and decides on a Specialized Status II downhill bicycle. The full suspension and quality components will be important for comfort and safety while traveling at high speeds. Ron has some technical skills and doesn't mind assembling an ebike from multiple vendor sources to make sure he can get exactly the components that he wants.

He chooses to go with a 48V system to achieve his desired high speed. He decides on a 48V10AH LiFePO4 battery from www.GoldenMotor.com (item number LFP-4810S) and a seat posted mounted rack (item number: RAK-002) from the same vendor site.

For a controller, he goes with a custom modded "9 FET 72V Infineon Brushless Controller " built by Edward Lyen and purchased at www.Lyen.com. He specifies that he'd like the controller preprogrammed by Lyen for 30A and to come with a Cycle Analyst plug pre-installed. This controller will provide him with more than enough power and has the highest quality electronics to put up with his use (and potential off-road abuse).

For a motor, Ron decides on the rear Nine Continent 2806 from www.ebikes.ca. Just the motor in a wheel without the rest of the kit is part number "M2806RD26". He also picks up a throttle from www.ebikes.ca (part number "T-HTwist") and a direct plug in Cycle Analyst to see his riding data (part number "CA-DP").

He'll need a serious torque arm if he wants to run such a powerful motor and controller, so he purchases one of DoctorBass's "Ultimate Torque Arms" from the EndlessSphere forum thread (http://endless-sphere.com/forums/viewtopic.php?f=31&t=29129&hilit=doctorbass+torque+arm). While waiting for his parts to arrive in the mail, he checks out a local scooter store and buys a full face motorcycle helmet, just in case. (It would be hard for him to impress women if he doesn't have any teeth left).

Ron's parts arrive and he gets to work assembling. He mounts the rack on his seat tube, extended over the rear wheel and also mounts his controller under the rack. His battery, controller, motor and throttle came from different sources so the connectors don't match, but he doesn't have too much trouble cutting off the connectors and simply soldering the wires together.

He has to play around with the phase and hall combinations a few times, but with tenacity he eventually finds the correct combination. He gets everything assembled and enjoys his first ride on a very powerful ebike.

His build results in a top speed of just about 30 mph (48 km/hr) and a range of about 15 miles (24 km), though if he slows down to about 20 mph he can eek out a range of closer to 22 miles (35 km). Ron uses his ebike for cruising on road as well as tearing it up off-road. Because he is traveling at such high speeds, he makes sure to bring his bike by the local bike shop at least once every other month for a tune up and brake check.

Total cost including S&H:

- Bicycle: $2400
- Battery & rack: $575
- Lyen controller: $110
- Motor, freewheel, throttle & Cycle Analyst: $460
- Helmet: $140

Total: $3,685

In closing...

I hope you enjoyed reading this ebook and I wish you the best of luck in building your own ebike. The first time you try something new, it always seems tougher, but just remember that in these pages you have all of the information that you need to successfully get through your first ebike build.

A good friend of mine and I were recently reminiscing about our first ebikes. He told me that ten years ago, back when ebikes were a fairly new commodity (and the parts weren't so simple to come by either) it took him 3 months to assemble his first ebike. Now, we can do one together in half an hour. It always seems tougher than it really is your first time. Stay with it!

The experiences of thousands of happy ebike builders show just how easy it can be, once you have a bit of experience and confidence, to get through any build, simple or custom, big or small, fast or slow. I look forward to hearing from you about your own ebike build, and I'd love to see pictures of your finished project! Keep in touch and remember to always ride safe!

Take care and best of luck to you!

--Micah

Photo Credits

Cover: Photo by Ohad Cadji; Page 1: Illustration by Kara Bocan; Page 15: photo by Micah Toll; Page 30: upper photo by Grand Canyon NPS via Flickr used under creative commons license, lower photo by BiohazardMan via Endless Sphere, used with permission; Page 33: photo by Micah Toll; Page 34: photo by Micah Toll; Page 35: photo by Micah Toll; Page 36: photo by Steve Ryan via Flickr used under creative commons license; Page 42: photo by Micah Toll; Page 48: photo by Micah Toll; Page 49: photo by Micah Toll; Page 50: photo by Micah Toll; Page 51: photo by Micah Toll; Page 56: photo and diagram by Micah Toll; Page 60: diagram by Micah Toll; Page 62: photo by Micah Toll; Page 64: photo by Micah Toll; Page 66: photo by Micah Toll; Page 68: photo by Micah Toll; Page 71: photo by Micah Toll; Page 72: image from ebikes.ca, used with permission; Page 73: image from ebikes.ca, used with permission; Page 83: original photo by Greg Morss, used under creative commons license; Page 92: image from ebikes.ca, used with permission; Page 93: photo by Ildar Sagdejev, via Wikipedia and used under free license; Page 97: screen capture from ebikes.ca, used with permission; Page 102: photos by Micah Toll; Page 115: photo by Micah Toll; Page 125: photo by Micah Toll; Page 127: photo set by Micah Toll; Page 128: photo by Micah Toll; Page 130: photo by Micah Toll; Page 131: photo by Micah Toll; Page 132: image from ebikes.ca, used with permission; Page 134: photo set by Micah Toll; Page 136: photo by Micah Toll; Page 137: photo by Micah Toll; Page 138: photo by Micah Toll; Page 140: photo by Micah Toll; Page 141: photo by Micah Toll; Page 144: photo by Micah Toll; Page 147: photo set by Micah Toll; Page 150: photo set by Micah Toll; Page 152: photo by Micah Toll; Page 156: photo set by Micah Toll; Page 158: photo set by Micah Toll; Page 161: photo set by Micah Toll; Page 173: chart by Micah Toll; Page 207: photo by Sapir Toll

Acknowledgements

Writing this ebook was enjoyable for me, largely due to the supportive group of people I had around me. First, I'd like to thank my wife, Sapir, for putting up with my many hours of sitting at the computer writing this book, and secondly for her patience when my mind gets lost on electric bikes for days, weeks or months at a time. Now that I've finished the book, I believe that she'll be happy to have my attention back (until the next project!).

I'd like to thank my parents, Ron and Kathy Toll, for encouraging me to pursue my engineering goals and to share the knowledge that I've obtained. Without their help and encouragement, I wouldn't be where I am in life today. Any success I've ever achieved has been, and will continue to be, a direct reflection of their good parenting.

I'd like to thank numerous mentors who have guided me in a number of ways over the last several years: Ms. Cheryl Paul and Dr. Howard Kuhn for supporting my engineering endeavors before I even knew what an ebike was; Ms. Gena Kovalcik and Dr. Eric Beckman of the Mascaro Center for Sustainable Innovation at the University of Pittsburgh, whose seemingly unconditional support and advice was and still remains invaluable to me today, and Mr. Andy Holmes of the Swanson Center for Product Innovation, who has probably done more than anyone other than my own father in shaping and guiding me as a person, an engineer and as a hard worker.

Last but quite certainly not least, my friends, Max Pless, Thorin Tobiassen and Gili all deserve a tremendous thanks for being there at different times throughout my electric bicycle career and helping me learn much of what I know about the subject today.

All of these people contributed in their own ways to helping me reach this point in my life as well as urging and encouraging me to continue even further toward attaining my goals and dreams, and I greatly appreciate this opportunity to acknowledge each of them.

Kickstarter Acknowledgements

All of the following people assisted in the publishing of this book by pledging financial support on Kickstarter.com. It is through their support that the paperback version of this ebook was able to be published. I owe each of them a sincere thank you, and I am glad to see their names honored here.

To my backers below: each and every one of you should know that without your support, you and countless others wouldn't be reading these words right now.

In no particular order (Ok you caught me, it's the order they answered the survey) I give thanks to: David Steinberg, John "Evernevermore" Scheib, Anonymous, Arnaud Stevins, ZeBadger, Tristan Ofield, Kenneth Ramey, That One Guy, Joeseph Allen Simon, Tyce Herrman, Steven Tom, Joseph Magin, Jon, Anonymous, Jay K. Jeffries, William Smith, David Mendoza Hernandez, Bob Waters, Jeff Mabbott, Darin Ramsey, Eden-Zeev Einav, James Hukill, Bill Ola Rasmussen, Britney Dupee, Radhakrishna Gorle, Scott Henshaw, David Bruner, Anonymous, Dan Morton, Charles Brian Peabody, Marco Ziegert, Geir Haukursson, Dr. Thomas J. Huff, Scott Owen, Dan-Andrei Danga, Jonas Atterbring, Chris Albrecht, Anonymous, Molly Pieri, Gary Saville, Budi Mulyo, Patrick J Murtaugh, Charles Wilhite, Ryan Hair, Izaura Maria Carelli, Martin Jackson, Wei Du, Derek Uttley, Nicholas Otto, Brad Sturm, Anonymous, Ron Mah, Anonymous,

Anonymous, Antonio Soto Miranda, Anonymous, The Soap Pedaler, Anonymous, Steve M. Fletcher, Fiid Williams, Steve Cooper, Geoff Cooper, Michele Lopez-Glynn, Robert Schlecht, Martin Evans, William Slattery, Arkane Loste, Chris Elliott, Brad Esopenko, Anonymous, Ido de Lepper, Anonymous, David Williams, Guilherme Martins, Triton Circonflexe, Veronica Wood, Andrea Mace, Colin O'Kelley, Claudio Kohn, Daniel Harden, Albert E. Wedworth, Shawn Higgins, Michael Brisson, Laura dwelly, Iker Agirre Unzueta, J.F. Allard, Morton Hoffman, Jerry L, John Kruse, Anonymous, Faye and Woody, Russell Reynnells, Jean Vermeiren, Marc Potvin, David Moberly, Anonymous, Andrew Sanders, Jim Kirk, Daniel Lemay, Nicholas R. Soika, Joel S Kondas, Howard Springsteen, Philip Egly, Jean-Claude Ragris, David Tanner, Gregory Bailey, Anonymous, Pedro Semeano, Anonymous, Thomas Essex, Kirk Bell, Hirokazu Sakamoto, Charlie Dick, Barbara Blackwell, Regina & Frederick Kitt, Jessica Ruby Radcliffe, Jeremy Rowland, Tylor M Balson, Judy Bonney, Anonymous, Anonymous, Anonymous, John Stalter, William J. Reiter, Anonymous, Quito Washington, Kevin Vaillant, Jacob Blumner, James Hartman, Doug Ittner, Anonymous, Anonymous, Praveen Tipirneni, Anonymous, Kevin Lepard, Rodrigo Perez, Tobi Johansen, John T Morrison, Linda Welker, Gary T Floyd, Pankaj Rajankar, Nick Politis, Gonzalo Saloma, Gary J Engelman, Tom Capiau, Bengt E. Josefsson, Shang Lin Tsai, Robert Bush-Kaufer, Thomas Hartmann, James Hartley, Anonymous, Anonymous, Pablo Gonzalez de Chaves Fernandez, Matej Kráľ, Volty, Julien Charbon,

Helen Collins, Don Albrecht, Christopher Nichols, Bill Murphy, Patrick Williams, Elad Karni, David Wood, Franny Jay, Leslie Shier, Paul Scarrone, Raymond Michaud, Eric Hervol, Vicki Wallis, Matt Bernard, Matthew Miller, Anonymous, Juan Carlos Santana, Tyler Krupicka, Ray Triana, Anonymous, Anonymous, Tom Dollmeyer, John W. Morehead, Steve Zeets, Anonymous, Paul Prawdiuk, Anonymous, Hugh Reynolds, Peter Thornton, Scott Chapman, Vincent Orton, Anonymous, Bolling Willse Sr, Anonymous, Christian Shuler, Damir Mestrovic, Mukesh Kacker, Yamazaki Yoshiharu, Gizmo Bourne, Anonymous, Shlomo Silverman, Shawn Jones, Adish Raza Padhani, Jeff Kinnebrew, Lisa Fuqua, Helenmarie DeLeon, Patrick Tse, Bryan Ferguson, Daniel Atri, Pete Steinke of Golectric Inc, Frank W Garrison, Jack Donnell, Lee Church, Steve Hofer, Anonymous, Lennart Vissers, Ioannis K. Erripis, David Cheeney, Guillermo Pérez, Justin L. Jenner, Anonymous, Anonymous, Ikaraam Ullah, Adam Rossi of Adam Solar Rides, Chris Bull, Paul Adams, Bill Cawthorne, Lori Berthelot, Adam R. Martin, Prashanth V, Joe Sanders, Michael E. McKinzy Sr., Anonymous, Anonymous, Anonymous, Vance McPhail, Edward Hudson, Wulf Oppenlaender, Fatih Ertekin, David Hinds, Derek Dyer, Ron and Kathy Toll, Prathep (Tep) Narula, Spencer T. Stewart, Francisco Valverde Manjon, Anonymous, Phillip Zaki, David Wheelock, Tres Randolph, Tayeb Habib, Warren Walsh, Lester Kober, Robert David Cyders, Gregmon, Gerald Dildine, Seth Goldenberg, David Grant, Ryan Carter, Frederick Schoch, Tom Spruill, Byron Young,

Daniel Reed, Bill Ang Siow Chen, Jonathon Richardson, and Ed Kopp.

About the Author

Micah Toll was born in rural Tennessee and graduated from the University of Pittsburgh's Swanson School of Engineering with a degree in Mechanical Engineering. While in Pennsylvania, Micah founded three startup companies, including an electric bicycle company. Micah and his wife Sapir live in Tel Aviv, where Micah has contracted and consulted with electric bicycle companies locally and internationally. He enjoys teaching others to build their own electric bicycles. Micah can be found riding the streets of Tel Aviv on his latest ebike creations.

What are you still doing here? The book's over. Go build an ebike!

Printed in Great Britain
by Amazon